THE SAILOR'S WEATHER GUIDE

BOOKS BY JEFF MARKELL

COASTAL NAVIGATION FOR THE SMALL-BOAT SAILOR
RESIDENTIAL WIRING
SOLAR TECHNOLOGY

THE SAILOR'S WEATHER GUIDE

BY JEFF MARKELL

SHERIDAN HOUSE

First paperback edition
published 1995 by
Sheridan House Inc.
145 Palisade Street
Dobbs Ferry, NY 10522

First published 1988 by W.W. Norton & Company, Inc.
Reissued with corrections 1995 by Sheridan House, Inc.

Printed in the United States of America

ISBN 0-924486-91-0

To SANDY *and to*
JOSE A. COLON *and* WILBUR SHIGIHARA
of the U. S. Weather Service

CONTENTS

PART III—MECHANICS OF WEATHER

LIST OF ILLUSTRATIONS

PREFACE

There have always been people who have thought it would be just dandy if we could control the weather. A few well-regarded experts have even tried, with somewhat less than complete success, to produce rain and subdue hurricanes by cloud seeding.

On the other hand, a considerable body of expert opinion is convinced that if we haven't already irreversibly ruined the climate of the earth, we are about to. We are polluting the air with industrial smoke, vehicle emissions, and aerosol propellants, destroying square miles of forests, filling square miles of tidal marshes, damming rivers, plowing up grasslands, irrigating deserts, and generally defacing the planet.

Certainly on a local basis there can be little doubt that human activity has produced some very serious adverse effects on the weather. The recurring smogs in London, Los Angeles and elsewhere, and the disastrous effects of acid rain on lakes and forests in Canada are examples.

However, the forces involved in even such a comparatively small weather system as a local thundershower, let alone an average cyclonic storm covering several thousand square miles, are truly colossal. Any sailor who has weathered a storm at sea knows how awesome these forces are. The impertinence of people to think we really can tame and control them is rather absurd. However, we have made giant strides, particularly in recent years, in the direction of understanding how these forces operate. By understanding the hows and whys of weather we can take the best advantage of good conditions, and avoid a great many difficulties.

After sailing on a merchant ship through a hurricane north of Cuba, and on other boats through several winter storms in the North Atlantic, I've made my present vessel a *pleasure* boat. If it's not going to be a pleasure, I'm not going out! One way of insuring that it will be pleasurable is to keep track of the weather both before I leave, and while under way.

You too can avoid most of the discomfort and unpleasantness of bad weather, and steer clear of many dangerous situations as well. Simply make use of the sources of weather information available to you both before and after departure, and keep a "weather eye" out while under way.

INTRODUCTION

The sailor, whether on an ocean liner or an 8-foot pram, has placed himself in the beautiful world that exists only between the wind and the water. In fair weather this can be a particularly wonderful place; in foul weather it can be quite the opposite. When you are not sailing under orders from the military or from a commercial owner, you are at liberty to put to sea only when conditions are pleasant, and stay in port when they are not. It is the purpose of this book to help you do that.

The U.S. Weather Service continually collects weather information both through its own sources and from others all over the planet. As this information is entered into the computer banks at weather headquarters the weather maps and predictions being issued are changed constantly with the moving weather systems.

Only a tiny part of the total planetary picture being maintained at weather headquarters applies to your present location, and to where you want to go. The most reliable, and timely, information on your immediate area is obtained directly, by radio or telephone, from the Weather Service office nearest you. The media (television, commercial radio, and newspapers) all get their information from Weather Service as well, but there is a time lag that may be considerable before it reaches you. Part I of this book deals with both the collection and distribution of weather information along with some thoughts on evaluating timeliness and accuracy.

Part II deals with several matters the small boat sailor should be aware of in dealing with the weather. Of primary importance is maintaining a good weather watch on your own vessel. This requires a few simple instruments. Weather systems do not always develop and move quite as the Weather Service expects. A front they expect to arrive nine or ten PM tonight might pick up speed and intensity and get here at one PM instead. Keeping track of cloud, pressure, and wind changes will alert you to the fact that it's moving in early.

The sailor encounters distinct and different seasonal weather patterns in different parts of the country. These result from several factors such as latitude, geographical contours, major planetary wind patterns, and ocean currents and temperatures. The discussion of usual seasonal weather in various areas will help you plan for normal conditions, and alert you to abnormal ones.

Fog is one of the sailor's worst enemies. The formation and dissipation of the several different types the sailor encounters are described, as well as the frequency of fog in various areas of the country during the different seasons.

The sailor needs constantly to be aware of the profound effects of the atmosphere on the oceans, as well as the effects of the oceans on the atmosphere. Tremendous interchanges of heat and moisture are constantly in progress between the two. The winds, in addition to causing waves, drive most ocean currents. Conversely the maritime air masses that influence so much of our weather result from the influence of the oceans on large blocks of air that linger over them for extended periods.

Part III explains the basic mechanics of weather systems, how they form, move, and dissipate. Clouds, wind direction, temperature and atmospheric pressure are the major indicators of what is presently happening, and what is likely to happen in the near future.

Violent storms follow very definite patterns as they develop, rage, and finally fade away. These patterns differ with the type of storm. Some are small-area, high-intensity storms such as thunderstorms, tornados, and waterspouts. The larger area warm core tropical storms, hurricanes and typhoons differ sharply from the huge temperate zone cyclonic storms.

Driven by colossal forces we have finally begun to comprehend but not control, giant weather systems form and move through our atmosphere. The major systems are then greatly modified by both atmospheric and surface factors. Making a minimal effort to understand and keep track of what is happening to the weather around us makes sailing far more pleasant, and much safer as well.

For help in compiling this weather guide I should like to thank the National Oceanic and Atmospheric Administration, and certain individuals in that organization. Specifically, my thanks to Mr. David B. Gilhousen of NOAA Data Buoy Center, and Mrs. J. C. David at NOAA Headquarters for illustrative material. Also Mr. Dan Atkin, Meteorologist, U.S. Weather Service, San Diego, kindly sat for the photo to illustrate operation of the AFOS computer system.

Thanks also to Sitex Corp, Weems and Plath, and Taiyo Musen Co. Ltd. for illustrative material.

Most of all my profound thanks to Mr. Wilbur Shigihara, Meteorologist in Charge, U.S. Weather Service, San Diego. He kindly went over the entire manuscript to root out any technical errors or misstatements of fact and contributed many helpful suggestions along the way.

PART I

WEATHER INFORMATION

1

SOURCES AND DISTRIBUTION OF WEATHER INFORMATION

Since earliest times sailors have been alternately delighted or infuriated, calmed or terrified, pleased or frustrated by the continually changing patterns of the weather. Living between the wind and the sea, the sailor must be particularly sensitive to weather changes. Anything that will help him foresee the sort of weather brewing beyond the horizon is more than welcome. If good weather is on the way he can relax. If foul weather is in the offing he can prepare for it, and would certainly be most unwise if he fails to do so.

EARLY WEATHER LORE—PROVERBS AND SAYINGS

Old time sailors, and farmers as well, had a wealth of weather sayings based very simply on years of observation and experience. A good many of these, under scientific analysis, have proven to be total nonsense. However, many others have very solid scientific explanations as we shall see.

Long ago there was an amusing old sailing direction for the trip from Boston to Cuba. It said, *"Sail south until the butter melts, then turn east to Cuba."* In July or August Boston is often hot enough to melt the butter before you leave the dock! However, here are some equally old sayings that prove to be firmly based on scientific fact:

Red sky at night, sailor's delight
Red sky in the morning, sailor's warning

Mackerel sky, twelve hours dry.

Rainbow in the morning, sailors take warning.

At sea with a low and falling glass, *[barometer]*
Soundly sleeps the careless ass;
Only when it's high and risin',
Safety rests the careful wise'un.

Thunder in spring, cold will bring.

Bright moonlight nights have the heaviest frosts.

A halo round the moon, foul weather soon.

Long foretold, long last,
Short notice, soon past.

From time to time, as you proceed through the book, look back at these old sayings. One by one, the explanations will become clear as you gain understanding of the mechanisms involved in the changing weather.

EARLY WEATHER OBSERVATIONS

Weather lore in the form of sayings such as those noted above became part of the oral tradition of prehistoric hunters, farmers, and mariners long before the development of writing.

Starting as early as 3000 BC, observations of this type were preserved in written form on clay tablets by the Babylonians. By 3000 BC the Chinese, as well, were foretelling the seasons, and their related weather changes by observations of the stars.

Peoples of the ancient world generally considered what we know to be natural phenomena as manifestations of supernatural powers. Thus the great Greek epics such as Homer's *Iliad* and *Odyssey* explained the changing moods of the weather as actions of the gods. Odysseus's stormy wanderings are attributed to the anger of mighty Poseidon, lord of the sea. The fair wind that finally brought him home was provided by the sympathetic "grey-eyed Athene."

Astrometeorology, the prediction of weather based on the movements of planets and stars, dates back to the Babylonians. It reached great heights of popularity in Europe during the later part of the Middle Ages. However, what may surprise you is that it continued to be highly respected until well into the seventeenth century! At that time the beginnings of scientific meteorology—the study of weather as we know it—were laid down.

Actually, by the time of Aristotle back in the middle of the fourth century BC, a far more scientific approach to meteorology had been started, but, with time, some parts were forgotten and others garbled. Aristotle's book, *Meteorologica,* became the authority on the subject for the next 2,000 years along with varying admixtures of astrometeorology and outright nonsense.

THE START OF SCIENTIFIC WEATHER OBSERVATIONS

Systematic weather observations began in the 1600s after the invention of the thermometer by Galileo, and the barometer (Fig. 1-1) by Torricelli. Pascal discovered that barometric pressure drops with altitude. Boyle, Charles, and Gay-Lussac later worked out the relationships between air pressure, temperature, and volume, that will be discussed in Chapter 8. Later yet these relationships were found to explain the vertical and horizontal movements of air masses that will be explained in Chapter 10.

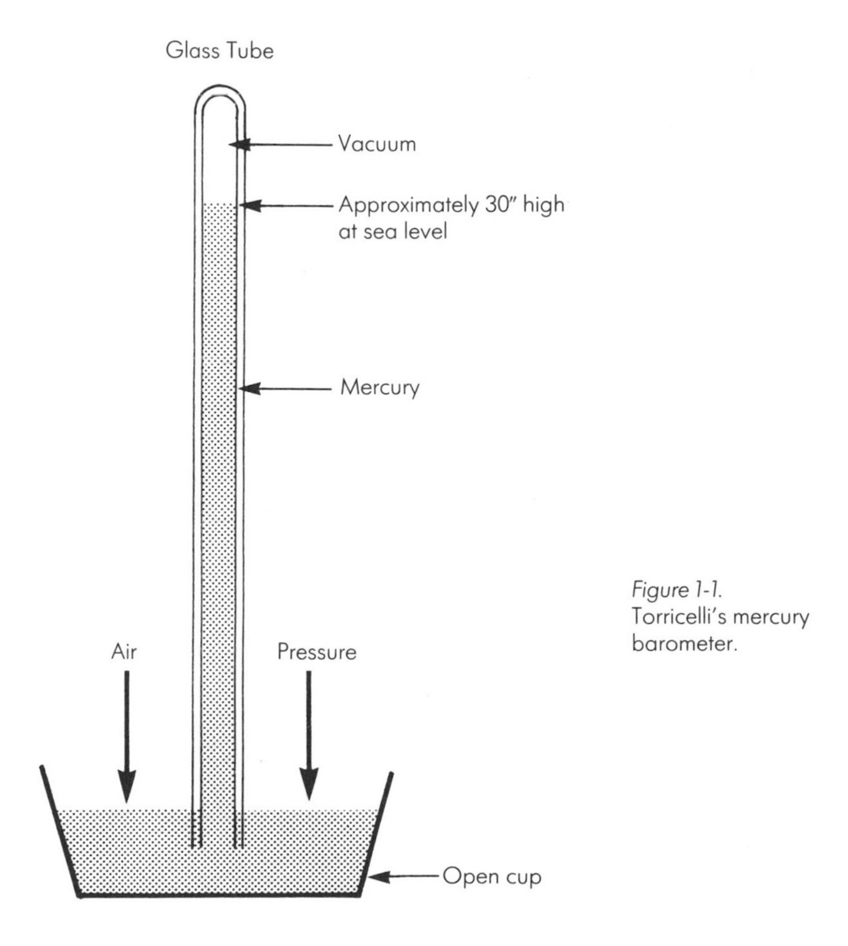

Figure 1-1. Torricelli's mercury barometer.

Much information leading to the present understanding of the basic air circulation patterns of the planet began to emerge from the data logged by voyaging sea captains. Captain William Dampier's book, *A New Voyage Round the World* (1697) was long a standard reference, and spurred many others to collect similar types of data.

Edmund Halley is best known for the comet bearing his name that returns every 80 years of so. Not so well known is the fact that he made several sea voyages toward the end of the 1600s resulting in the publication of very accurate charts of the earth's trade wind and monsoon wind circulations. His work was soon expanded and improved on by George Hadley whose explanation of the basic air circulation systems in the tropics—meaning from about 30° N to 30° S latitude—is still at the core of our present understanding of weather in those regions.

Charles Darwin and the controversy over his theory of evolution has long since overshadowed the importance of his captain on the HMS *Beagle,* Robert Fitzroy. During the famous voyage on which Darwin was his passenger, Fitzroy collected in his logs a vast wealth of weather information on the world's oceans. Later, as Admiral Robert Fitzroy, he became Britain's first official "weather man" becoming head of the Meteorological Department of the Board of Trade.

Ben Franklin, a man whose innumerable interests included flying kites in thunderstorms, was obviously intensely interested n the weather, and astronomy as well. In 1743 an eclipse of the moon was predicted for 9:00 PM on October 21 at Philadelphia. At the appointed hour, Franklin's view was obscured by a storm. He later heard that his brother in Boston, 384 miles to the northeast, had an excellent view of the eclipse at 9:00 PM. and that stormy weather arrived there at about 11:00 PM. After collecting information about events that had occurred in the area between the two cities, and elsewhere in the colonies, Franklin was able to trace the movement of the storm all the way from Georgia northeastward into New England. This constituted the first systematic study of the movement of a single weather system ever to be made in America.

COLLECTION OF WEATHER INFORMATION

Today we have a vast network of Weather Service land, sea, air, and satellite observation stations reporting on weather conditions all over the planet. In addition, commercial aircraft and ships at sea are still a vital part of this immense weather information system.

As Ben Franklin and others discovered, the weather system affecting your area today was somewhere else yesterday, and will continue on its way by tomorrow. Assembling and mapping information of current weather activity at a very large number of locations, as well as how it is moving and changing at those locations, produces an overall picture of the major planetary weather systems and how they develop and move. Analysis of the general picture of weather movements over large areas makes it possible to predict local changes at specific points within those areas with reasonable accuracy.

The National Weather Service has developed an extremely large-scale observation network to collect, correlate, and interpret weather information. Their results are made available to the sailor in various ways. *Caution:* No amount of information from other sources can, or should, replace the skilled eye of the mariner who must keep watch over conditions in his immediate vicinity.

SOURCES OF WEATHER INFORMATION

The comprehensive view of global weather that can be obtained from space, first became available in 1960 with the launching of *Tiros I* by the U.S. The following year the World Meteorological Organization (WMO), in response to a request from the United Nations, began to form an international global atmospheric research and meteorlogical program. Two years later, these efforts crystalized in the organization of the World Weather Watch (WWW).

The WMO has set standards for the uniform measuring and recording of weather data allowing meaningful international exchanges of such information that were not previously possible. Uniform numerical codes for rapid transmission of this information have also been developed.

The WWW has investigated many weather phenomena and various influential meteorological factors of international and world significance. For example, they have researched the atmospheric effects of such major events as the eruptions of Mount Agung and Mount Saint Helens, as well as the recurrences of the warm "El Nino" current in the Pacific Ocean.

By now, surface observations are made daily at standard hours at approximately 9,200 land stations in 150 nations, plus perhaps 7,000 ships as sea. About 850 stations make upper atmosphere observations. In addition observations are made from 5 geostationary satellites, and 5 more in polar orbits, with more on the way.

These satellites transmit pictures, such as those often seen on TV news programs. They show the major cloud formations associated

Figure 1-2. Satellite photo showing large weather disturbance in the Gulf of Alaska heading toward the western United States. Courtesy: U.S. Weather Service.

with large storm systems as well as the large clear areas where fine weather prevails (Fig. 1-2). By comparing pictures taken some hours apart, the direction and rate of movement of these large weather systems can be determined. As we shall see in Chapters 10 and 11, based on a careful record of past movements, the future movements of many storms can be predicted with considerable precision. Unfortunately, as we shall see in Chapter 12, all too often the movements of the most violent storms turn out to be the most difficult to predict.

LAND-BASED WEATHER STATIONS

Satellite pictures provide an excellent view of the size and activity of large-scale weather systems. However, precisely what is going on inside those systems is largely a matter of conjecture without detailed information provided through a network of ground observation stations.

Consequently, a very extensive network of such ground stations exists. These stations monitor temperature, barometric pressure, relative humidity, type and extent of cloud cover, wind velocity and direction, precipitation (rain, snow, sleet), and visibility. This information is regularly transmitted from all over the country, and the world, to National Weather Service headquarters in Washington as well as two other world meteorological centers located in Moscow and Melbourne.

The standard hours for observations are 0000Z (12 midnight), and 1200Z (12 noon). "Z" you may recall, is the abreviation for "Zulu" meaning *Greenwich Mean Time*. This is the same Zulu time normally used as the basis for the observations and calculations used in celestial navigation.

In addition to making surface observations, instrumented balloons are released from many ground stations carrying radio transmitters called *"radiosondes"* (Fig. 1-3) from which those stations collect information on conditions at various altitudes. When correlated, these reports produce a three-dimensional picture of atmospheric conditions. This three-dimensional picture is further enhanced by regularly scheduled wind-finding flights reaching up to altitudes as high as 100,000 feet. These wind-finding flights also supplement the temperature, pressure, and humidity readings from the radiosondes.

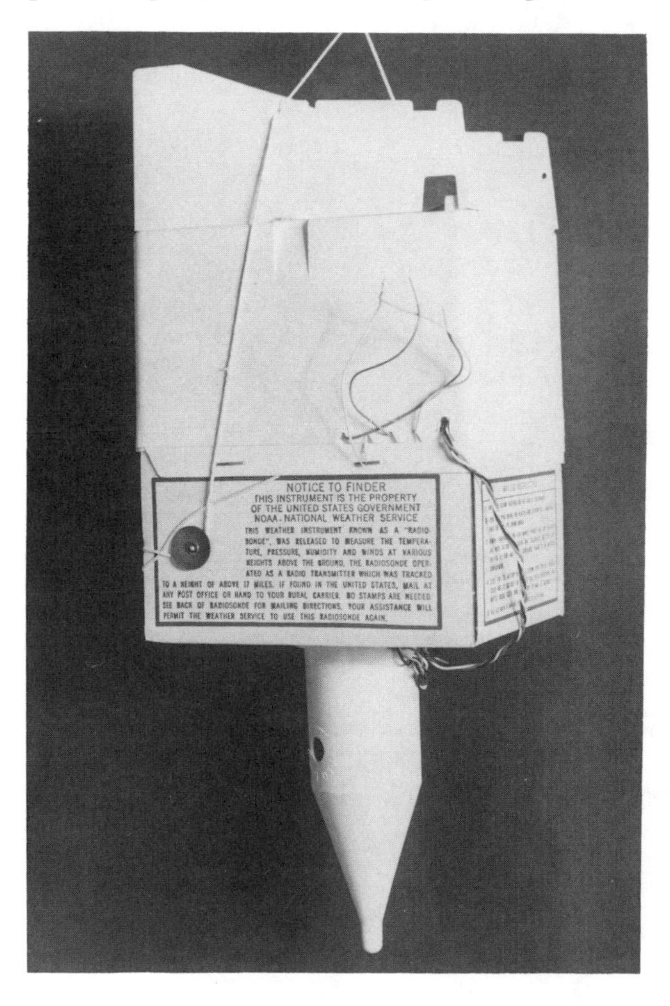

Figure 1-3. Radiosonde. Instrument package that is carried to great heights by balloon. While aloft, it transmits to the ground station temperature, humidity, pressure, and winds at various altitudes as it rises. Courtesy: National Oceanic and Atmospheric Administration.

In recent years, many manned coastal light stations and lightships have been replaced by automatic lights and Large Navigational Buoys (LNBs), or Superbuoys (Fig. 1-4). The personnel that formerly manned these positions routinely supplied weather observations to National Weather Service. The lack of those observations has constituted a significant void. To correct this situation, National Data Buoy Center is developing automated marine observing and reporting stations for use on LBNs as well as on the newly automated headland and island light stations.

Figure 1-4. United States Coast Guard Large Navigational Buoy (LNB) also called "Superbuoys." Courtesy: U.S. Coast Guard.

DISCUS AND NOMAD DATA BUOYS

The network of ground observation stations is fine as long as there is ground for them to stand on. However, as we sailors well know our planet is *mostly* water.

Consequently, another recent addition to the information gathering system of the Weather Service has been a group of large buoys placed offshore to collect and transmit weather information. These buoys are placed at sea by the national Data Buoy Center (NDBC) which is part of the National Oceanic and Atmospheric Administration (NOAA). Incidentally, National Ocean Survey (NOS), which prints the charts we use in U.S. waters, is another part of NOAA.

These buoys collect both weather and oceanographic data, and are of two types. One type is called *Discus* (Fig. 1-5). The Discus are large round buoys built in two sizes: 33 feet and 40 feet is diameter. The second type is boat-shaped, about 20 feet long, and is called *Nomad* (Fig. 1-6).

Some of these buoys are moored at specific locations in the Atlantic and Pacific Oceans, the Gulf of Mexico, and the Great Lakes.

Figure 1-5. "Discus" weather buoy. Courtesy: NOAA Data Buoy Center.

Figure 1-6. "Nomad" weather buoy. Courtesy: NOAA Data Buoy Center.

Others are adrift. The Discus type is used for the deep ocean moored buoys. Nomads are moored at nearshore locations and in the Great Lakes. Reports from the moored buoys include wind speed, wind direction, air temperature, barometric pressure, wave height, wave period, and sea surface temperature. The drifting buoys report their positions at sea in addition to all the same environmental data.

These data buoys greatly extend the geographical area from which the National Weather Service receives regular reports on surface conditions. In turn, this enlarged area of surface coverage results in much improved forecasting accuracy.

WAVE METERING STATIONS

Independent of the weather-data buoy system operated by NOAA, the U.S. Corps of Engineers operates a network of ocean *wave metering arrays.* These stations are located close to the coasts. Although placed and maintained by the Corps of Engineers they transmit information on coastal sea conditions such as wave height, period, and direction to the Weather Service.

AIRCRAFT AND SHIP REPORTS

Reports from a great many ships at sea further extend the geographic range of Weather Service knowledge and understanding of surface conditions. Unfortunately, since these ships are constantly moving, their coverage is necessarily spotty and irregular. However, taken all together they add invaluable information to the total picture.

The automated observing and reporting stations that now replace the old manned lightships and lighthouses are being modified for shipboard installation. In time this equipment, the Shipboard Environmental Data Acquisition System (SEAS), will be placed on a multitude of ships as a major supplement to satellite observations in areas of open ocean. Such reports have the advantage of not being influenced by variations in human perception and interpretation, or just plain human error!

Adding to both the geographic range and the three-dimensional picture are the many reports from aircraft. They operate at a wide range of different altitudes. Consequently, aircraft reports are particularly useful for filling in the three-dimensional view of atmospheric activities that is partly derived from radiosonde observations.

Both airplane pilots and ship's officers must become very precise weather observers as a part of their normal professional duties. Consequently, the Weather Service can and does place great reliance on the accuracy of their reports.

COORDINATION OF INFORMATION

All of the information produced by this entire elaborate collection system is funneled to the National Meteorological Center in Camp Springs, Maryland, via an immensely sophisticated and powerful system of interconnected computers. This system is called Automation of Field Operations and Services, or AFOS for short.

The computers of AFOS all over the country (Fig. 1-7) are linked by special telephone lines to each other and to Weather Service headquarters in Maryland. The observations taken at the multitude of outlying stations are fed via these telephone lines into the central computer banks at the National Meteorological Center (Fig. 1-8). There they are coordinated into a variety of weather maps and forecasts serving a number of different needs.

We sailors are not the only ones in need of accurate weather information. The farmer needs to plan planting, harvesting, and irrigation based on what sort of weather is expected. Construction companies will want to postpone pouring concrete if rain or freezing conditions are due. Airlines obviously need accurate weather information to guide their pilots safely.

Figure 1-7. Meteorologist Dan Atkin at the computer terminal in the San Diego office of the U.S. Weather Service.

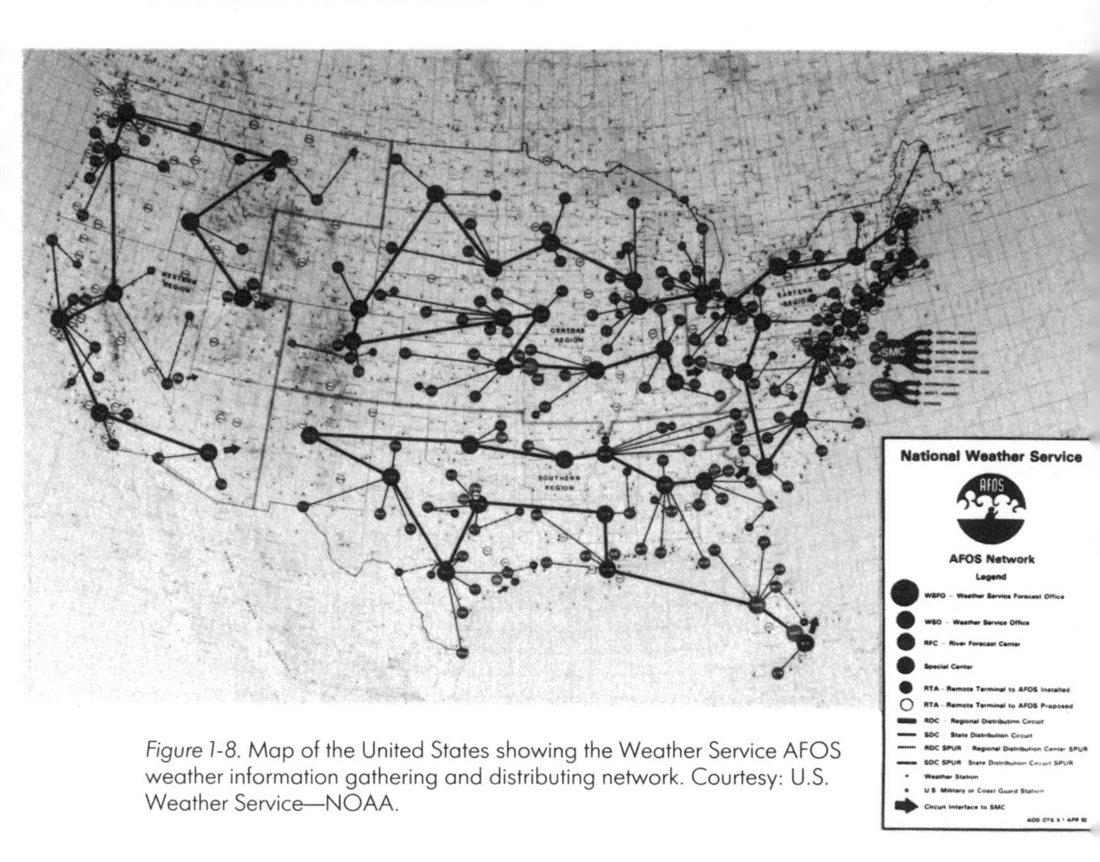

Figure 1-8. Map of the United States showing the Weather Service AFOS weather information gathering and distributing network. Courtesy: U.S. Weather Service—NOAA.

In response to the varying needs of commerce and industry, as well as those of marine interests, separate maps and forecasts are made for temperatures (Fig. 1-9), rainfall (Fig. 1-10), winds, and pressures. For aeronautical use maps of conditions at a variety of different altitudes are available. Combined surface forecasts and maps are also produced including all basic weather elements (Fig. 1-11). These combined maps are the basis of the newspaper and TV maps and reports.

All of these different kinds of maps and reports are fed back out to stations all over the country through the same AFOS network that transmits their raw observations in to headquarters. Also through AFOS, individual field stations can communicate with each other via State Distribution Circuits (SDC), and Regional Distribution Circuits (RDC) (see Fig. 1-8). This allows them to track developments of local importance through direct contact with each other.

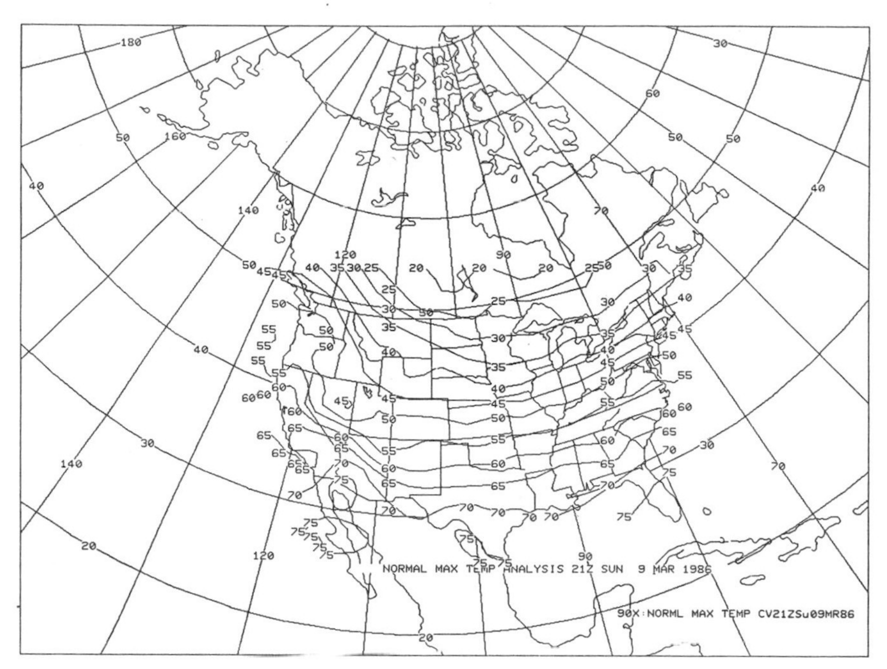

Figure 1-9. Printout of AFOS temperature map of the U.S. for 9 March 1986. Courtesy: U.S. Weather Service.

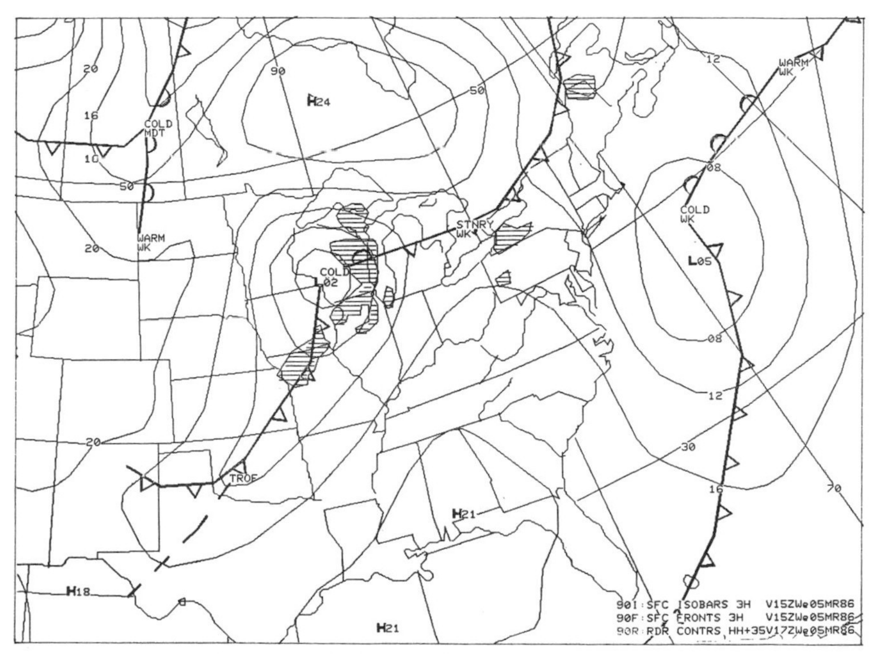

Figure 1-10. Printout of AFOS rainfall map of the eastern U.S. Courtesy: U.S. Weather Service.

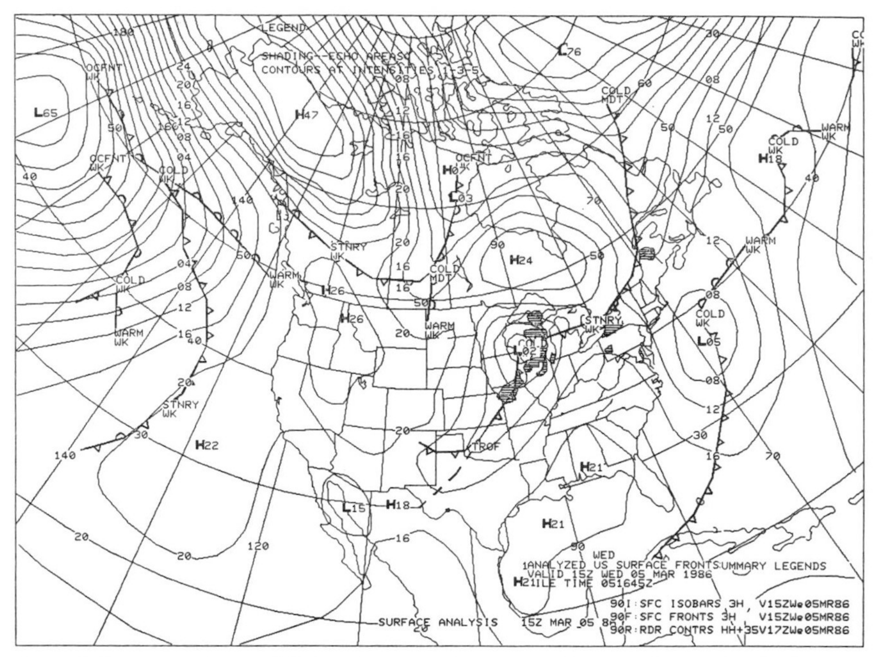

Figure 1-11. Printout of the AFOS surface weather map for Wednesday, 5 March 1986. Courtesy: U.S. Weather Service.

DIRECT LOCAL OBSERVATION

One of the first and most fundamental rules of good seamanship is to "maintain a sharp lookout at all times." This means that the competent skipper keeps his eye not only on the marine traffic moving in his area, but also on the sea, the sky, and the wind. He is constantly on the watch for changes in the weather or the sea conditions.

The Weather Service provides information on present conditions as well as forecasts for periods ranging from 12 hours to several days in advance. The areas covered vary from nationwide, to regional, to local in scope. A "local" forecast, however, covers at least an area of several hundred square miles. Within an area of that size, considerable localized differences will normally occur. For example, the Weather Service office in New York may issue a forecast for today saying "scattered thunderstorms" in the New York / Long Island area. That will cover a span from about Block Island to Sandy Hook—a distance of well over 100 miles in length. In width it will mean a band 40 to 50 miles across covering from the Atlantic coastal waters of Long Island across the Long Island Sound to the Connecticut shore.

Figure 1-12. Cumulus building upward; possible thunderhead.

A thunderstorm, as we shall see in Chapter 12, is very intense but very small in area. The above prediction, however, covers a total of 400 to 500 square miles! Anywhere within that space you may encounter thunderstorms, or it may remain a perfectly clear day with no weather problems at all. But in light of that prediction, you'd better keep a careful watch for cumulus clouds with towering vertical development (Fig. 1-12) that could become thunderheads. The life cycle of a typical thunderstorm is described in detail in Chapter 12 so when such a warning is issued in your locality you will know exactly what to look for.

Your local forecast may indicate that a *warm front* or a *cold front* is due to pass through the region during the day. Descriptions of the typical cloud sequences and wind changes that accompany the passage of such frontal systems are found in Chapter 10.

Compare what you are actually seeing with the cloud, wind, and temperature changes typical of the conditions predicted by Weather Service. If you are seeing what you should see, relax; you have a good handle on what is coming. If you are not seeing the right changes, the expected weather system has perhaps either stalled, or changed direction. Try your radio for a later forecast, and check your own instruments. A competent skipper does not allow himself to be caught unprepared by weather changes.

WEATHER INSTRUMENTS

The final source of accurate weather information at your precise location is the readings from your own instruments combined with your practiced interpretation of what you see. On a small boat an elaborate and expensive set of weather instruments is unnecessary unless you're making a hobby of meteorology. But to have none at all is unwise unless your boat is *really* tiny.

Certainly, if you are fishing from a 14 or 16 foot open outboard you are not going to have any space at all for weather instruments aboard. However, even on boats that small it is advisable to carry a self-contained handheld VHF radio (Fig. 1-13) not only for weather

Figure 1-13. Fifty channel handheld VHF-FM radio. Courtesy: Apelco division of Raytheon Marine Co.

information, but also as a standard piece of communications equipment for safety reasons. These units are available with models varying from 6 channels up to 55 channels, which is just about all the channels there are. On the smallest, the 6-channel type, two of the channels will be taken up by Channel 16, the required initial calling and distress frequency, and Channel 6, the intership safety channel. Along the coasts or in the Great Lakes, a third channel should be 22A, the Coast Guard working frequency. That leaves three still available. Assign one of those to weather, either W-1, W-2, or W-3, whichever one broadcasts coverage of the area in which you are sailing. The last two can then be used for the most popular intership channels among sailors in your area.

On somewhat larger boats with a cabin, a modest weather instrumentation is desirable. Any of us can tell when it feels warmer or cooler, but not many can gauge the actual extent of a temperature change at all accurately without a *thermometer* (see Fig. 3-3). Finding an adequate and inexpensive thermometer should present no problem.

Few, except some with arthritis, can feel changes in atmospheric pressure. Such changes are extremely important indicators of future weather. To follow pressure changes we must have a *barometer*. Barometers of the aneroid type (see Fig. 3-1) suitable for marine use vary in cost from a few dollars to several hundred. For our purposes a relatively inexpensive instrument will do. Accuracy increases with cost, but as we shall see in Chapter 3, we have no need for a very high degree of accuracy in the actual reading. We are primarily concerned with changes. We want to know, over time, the direction, the magnitude, and the speed of any changes in the atmospheric pressure. This does not require an expensive or highly accurate instrument.

A good many sailboat sailors, and a few motorboat operators as well, become remarkably accurate at estimating wind speed. Merely the feel of the wind on the face, and the look of the ripples on the water surface are enough for some. On a sailboat, the angle of heel and the speed of movement through the water tell the experienced skipper a great deal. A masthead *anemometer* (Fig. 1-14), gives a very accurate wind speed reading, but it's also quite expensive. Fortunately, a couple of reasonably priced handheld wind speed indicators (Fig. 1-15) are available that are accurate enough for our purposes. If our handheld unit tells us the wind speed is 25 knots, and it's actually 28, no matter. In either case it is time to think about reefing sails on a small sailboat.

Figure 1–14. Sailboat masthead anemometer.

Figure 1-15. Handheld wind speed indicator.

When space allows for one more small onboard instrument, a *sling psychrometer* is desirable (see Fig. 3-11). With this instrument you are able to determine relative humidity. Watching changes in humidity can often be the key to predicting one of the sailor's major weather hazards, fog.

With a fair bit of practice, your own observations, supplemented by a couple of instrument readings will enable you to look ahead at the weather in your immediate vicinity for limited periods of time with confidence that your predictions will prove reasonably close to the truth. Based on their vastly greater store of information the Weather Service is able to project a good deal further ahead, and with much greater accuracy than we amateurs can. Their predictions are made available through several channels. Depending on one's needs and interests, some of these predictions are more complete and detailed than others. Some cover large geographic areas, some small, and some are more current than others depending on how they get from Weather Service to you.

2

PUBLIC ACCESS TO WEATHER REPORTS

The collection, assembly, tabulation, and interpretation of weather information is a neverending activity at the Weather Service. It goes on 24 hours a day, every day of every year. As we have already seen, very advanced and sophisticated methods are then used to develop accurate weather predictions from this information.

However, no matter how accurate and detailed the information and predictions developed by the Weather Service may be, they are useless to you unless you have timely access to them. It is not going to help you one bit to discover today that Weather Service accurately predicted the foul weather you suffered through yesterday, but were not prepared because you didn't know of the prediction.

The National Weather Service makes its information available to the public in a variety of ways, both direct and indirect. Consequently, if you didn't know they were predicting foul weather yesterday it's your own darn fault! You could have known, and you should have known!

CONTINUOUS VHF MARINE WEATHER BROADCASTS

One of the normal functions of a great many of the local National Weather Service offices is the operation of a VHF radio broadcast station. These radio stations broadcast nothing but weather information continuously, 24 hours a day with no commercials!

The continuous broadcast service is intended primarily for both marine and agricultural interests. Transmissions are on the VHF-FM Marine Band. The broadcast from a typical coastal station consists of several parts:

1. Weather synopsis and predictions for coastal and nearby inland areas.

The general synopsis describes the major weather systems moving in the area including any fronts or air masses that have just passed, those that are currently passing, and any that are expected during the next 24 to 36 hours. Based on that synopsis, weather predictions for the general area are given covering today, tonight, and tomorrow. This will include expected wind directions and velocities, type and extent of cloud cover, coming precipitation (if expected), anticipated high and low temperatures both coastal and inland, and storm, high wind, heavy seas, highway icing, or other warnings.

Figure 2-1. United States Atlantic and Gulf Coast coastal offshore weather forecast areas. Courtesy: Worldwide Marine Weather Broadcasts—NOAA.

2. Agricultural forecast and warnings.
This portion deals specifically with field working conditions for farmers. It also covers sun and temperature conditions for local crops as well as warnings regarding wind, rain, or frost dangers to crops.

3. Marine forecast and warnings.
The marine forecast is made in three parts. The first deals with the immediate coastal and harbor areas. The second includes nearby coastal waters and up to about 60 miles offshore. The limits of the various U.S. coastal areas for which VHF forecasts are transmitted are shown in Figures 2-1 and 2-2. Figure 2-1 includes the East and Gulf coasts; Figure 2-2 covers the Pacific Coast.

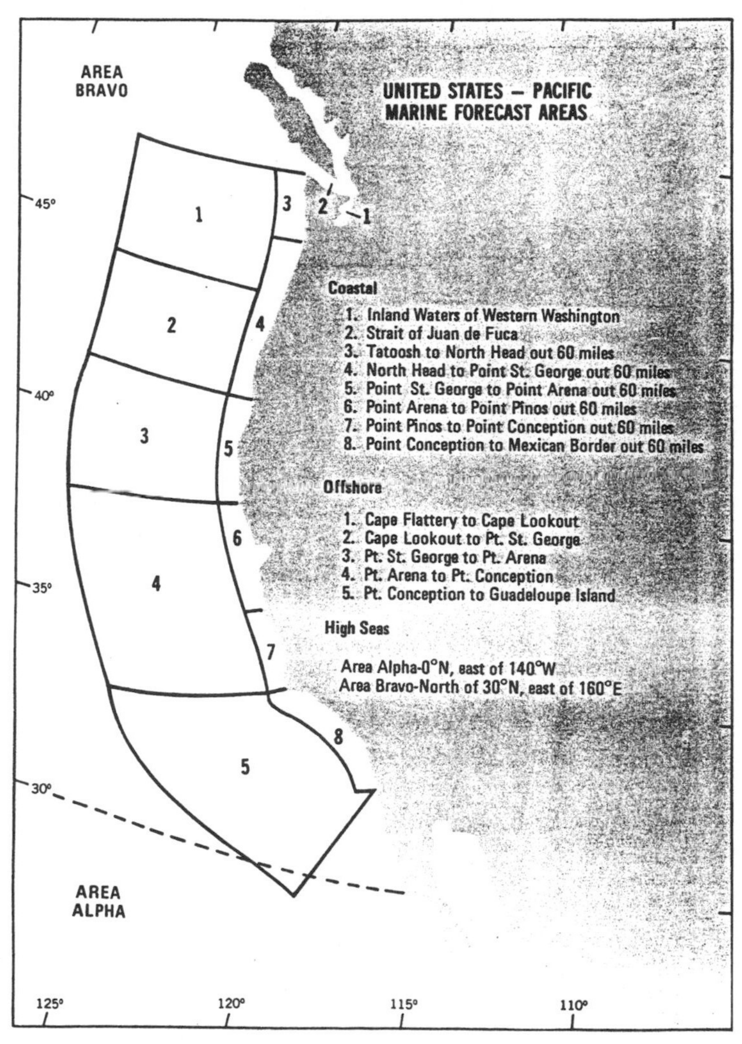

Figure 2-2. United States Pacific Coast coastal and offshore weather forecast areas. Courtesy: Worldwide Marine Weather Broadcasts—NOAA.

The third part of the marine forecast section covers offshore waters. Along the U.S. Pacific Coast this means from 60 to about 250 miles offshore as shown in Figure 2-2. Along the U.S. Atlantic and Gulf coasts, offshore distances vary as shown in Figure 2-1.

All three forecasts include expected wind directions and velocities, height of swells, high and low temperatures, extent and types of clouds, fog (if likely), rain or other precipitation, and any small craft or gale warnings that may be in effect or are anticipated.

4. Coastal weather observations.
This section of the broadcast lists current weather conditions as observed at a number of nearby coastal locations. The aspects noted include distance of visibility, wind direction and velocity, sea conditions, swell height and period, water temperature, air temperature, and barometric pressure.

VHF WEATHER BROADCAST CHANNELS

At present in the U.S. three VHF radiotelephone channels are used for continuous Weather Service broadcasts. 162.55 MHz is designated as W-1, or Weather 1, 162.40 MHz is Weather 2 (W-2), and 162.475 is Weather 3 (W-3). You will find these channels so marked on some VHF radios. On others, particularly the new sets covering 55 or more channels, they are marked simply 01, 02, and 03. The older 10 or 12 channel sets are likely to receive only one of the weather channels. Newer sets ranging from 24 channels on up usually have all three.

Depending on your operating range, you will want, if possible, to have crystals installed in your radio for the frequencies of all weather stations within that range if you do not already have them. As we shall see, at some locations more than one weather station can be received.

The cities where VHF weather stations are located on the U.S. Northeast and Great Lakes coasts are shown in Figure 2-3 along with their call letters and broadcast frequencies. In addition to the continuous VHF broadcasts of the Weather Service covering the Great Lakes area, Coast Guard also gives the following scheduled weather broadcasts on Channel 22:

CITY	STATION	TIMES
Buffalo	NMD-47	Every 3 hrs after 0255
Duluth	NOG-14	Every 3 hrs after 0135
Saulte St. Marie	NOG	Every 3 hrs after 0005

You will notice that as you move along any coastline, for example (Fig. 2-3) from Boston, to Providence, to New London, each station

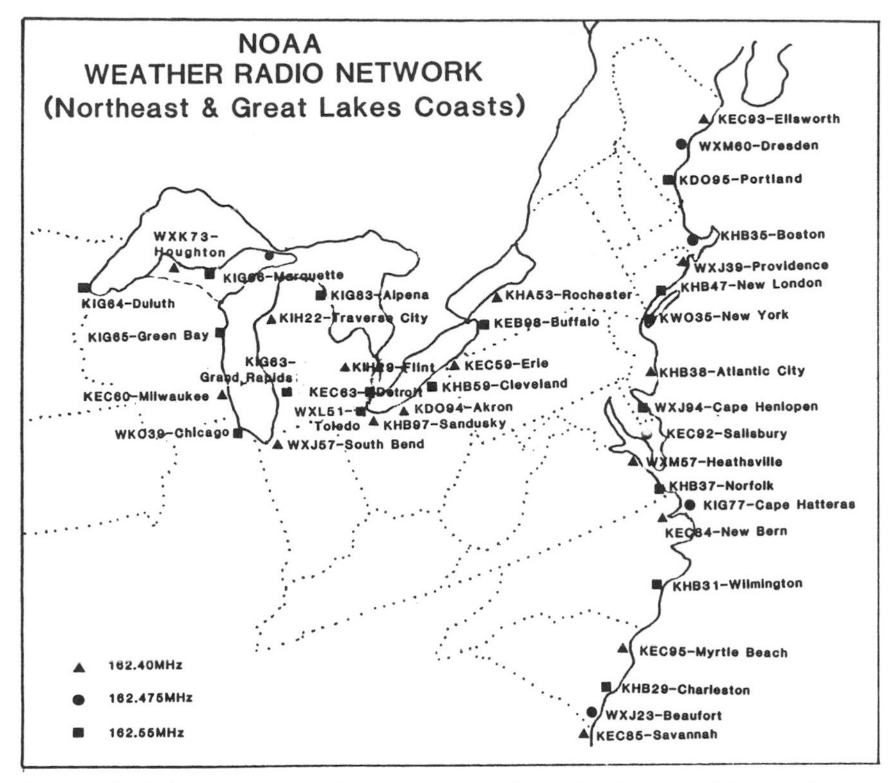

Figure 2-3. VHF continuous weather broadcast stations—Northeast and Great Lakes coasts. Courtesy: Worldwide Marine Weather Broadcasts—NOAA.

is on a different channel. This is because their reception ranges may overlap. There are many locations where you can receive both Boston and Providence, and some where you'll get New London as well. Since their ranges overlap, they must be assigned different channels to avoid interfering with each other.

VHF weather stations on the U.S. Southeast and Gulf coasts are shown in Figure 2-4 along with their call letters and broadcast frequencies. The Pacific Coast is similarly covered in Figure 2-5. Along the Pacific Coast, stations are located far enough apart that, except for the extreme northwest, only two frequencies are required to avoid interference (see Fig. 2-5). The spacing in both Alaska and Hawaii is again wide enough that only two channels are needed (Fig. 2-6).

Readers who operate along the Atlantic Coast as far north as Newfoundland, the St. Lawrence Seaway, or other parts of eastern Canada should know that Canada uses two additional weather channels in these areas. Channel 21 at 161.65 MHz is one, and Channel 83 at 161.775 MHz is the other (Fig. 2-7). In the Great Lakes, Canada has a number of VHF weather stations using those frequencies (Fig.

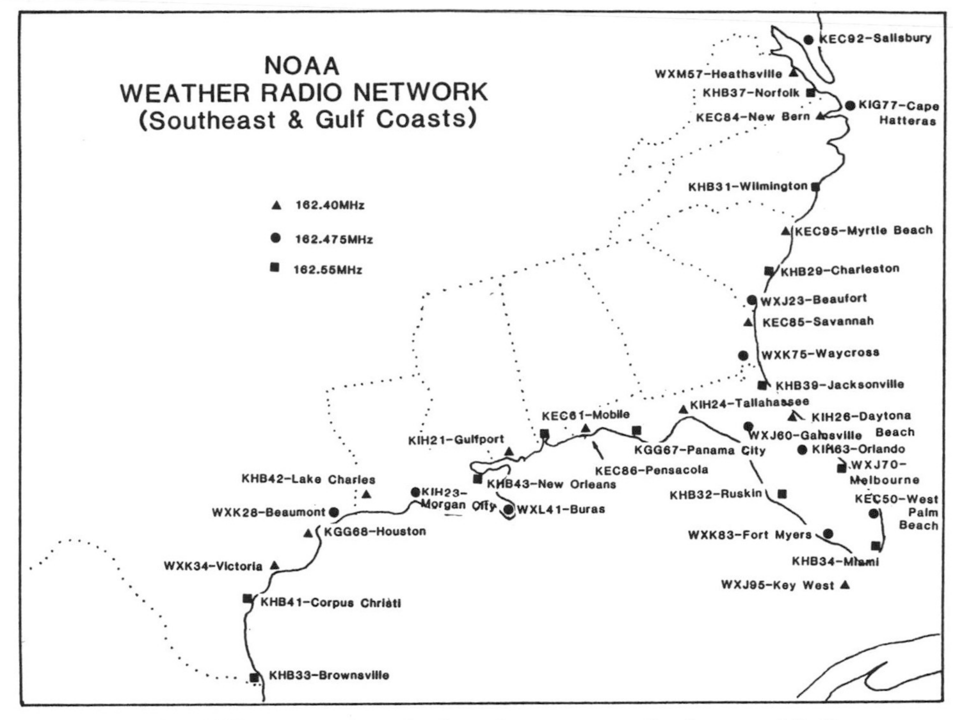

Figure 2-4. VHF continuous weather broadcast stations—Southeast and Gulf coasts. Courtesy Worldwide Marine Weather Broadcasts—NOAA.

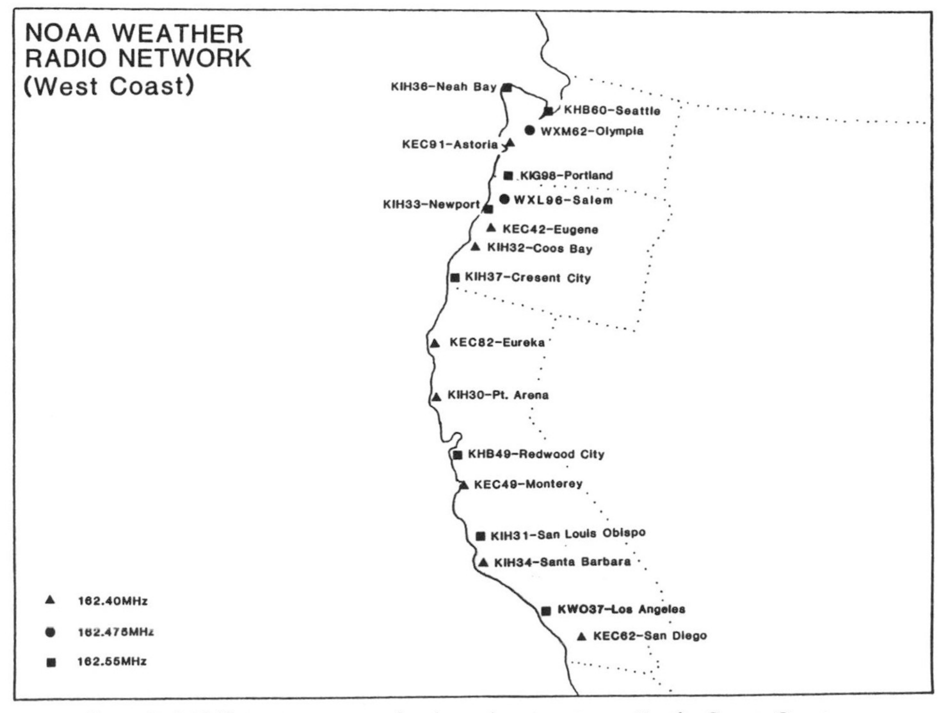

Figure 2-5. VHF continuous weather broadcast stations—Pacific Coast. Courtesy: Worldwide Marine Weather Broadcasts—NOAA.

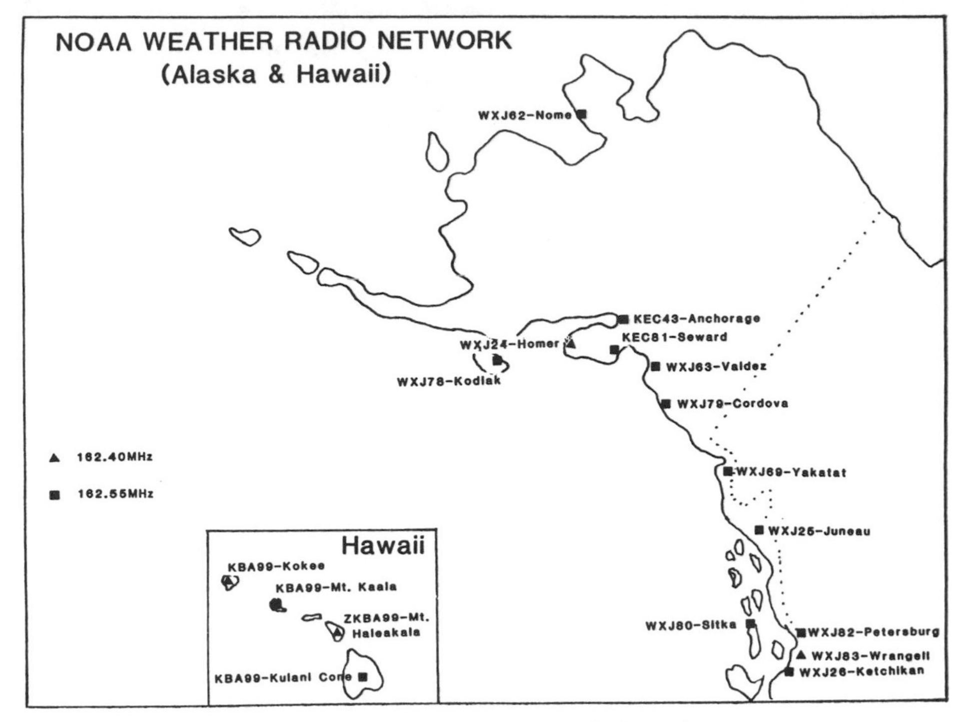

Figure 2-6. VHF continuous weather broadcast stations—Alaska and Hawaii coasts. Courtesy: Worldwide Marine Weather Broadcasts—NOAA.

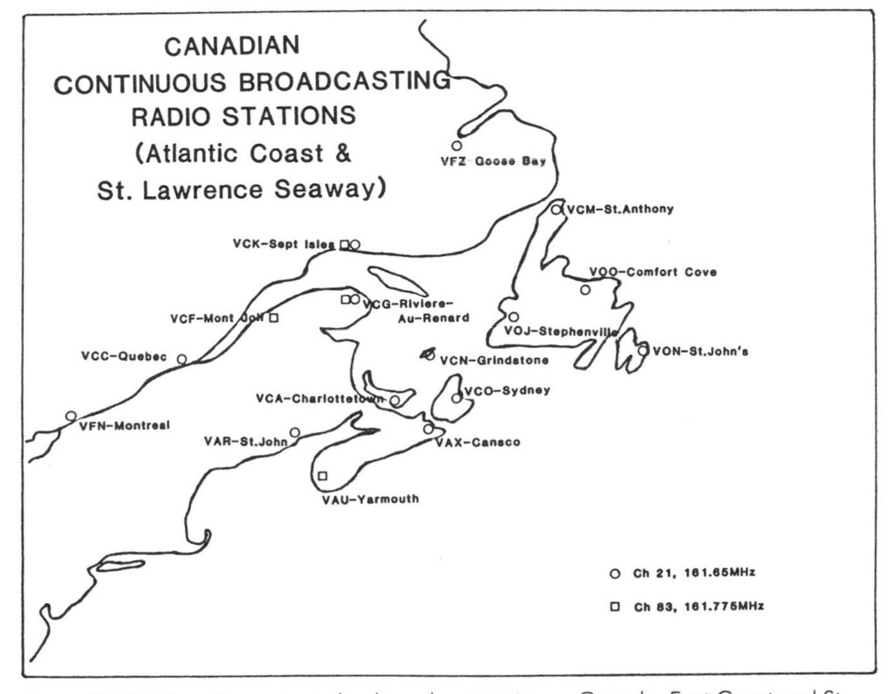

Figure 2-7. VHF continuous weather broadcast stations—Canada, East Coast and St. Lawrence Seaway. Courtesy: Worldwide Marine Weather Broadcasts—NOAA.

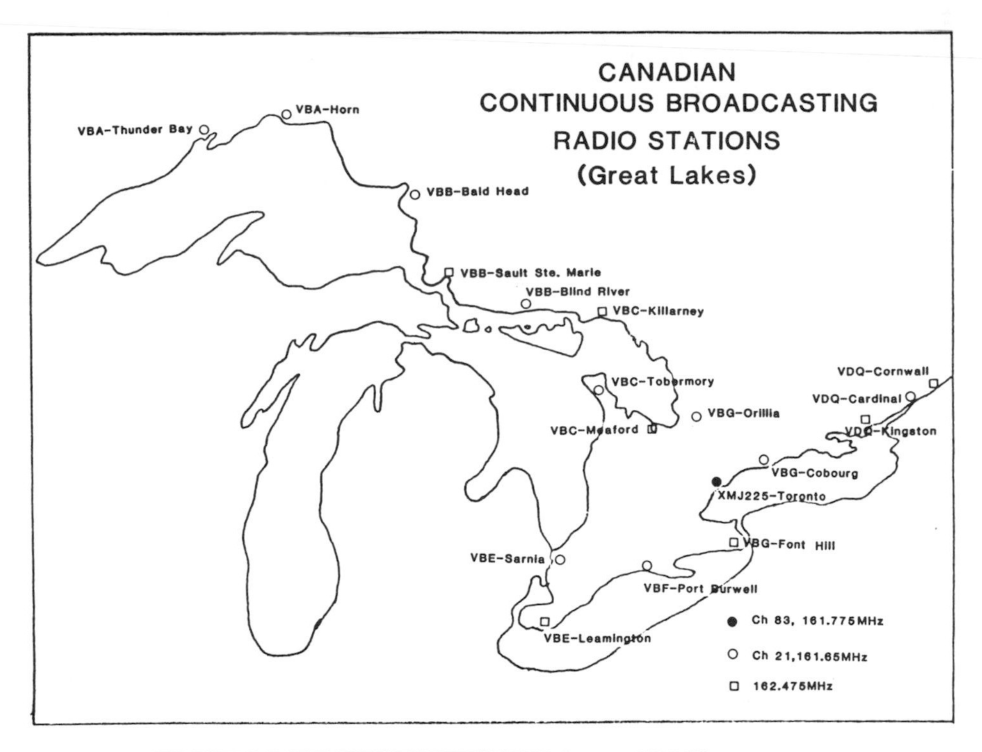

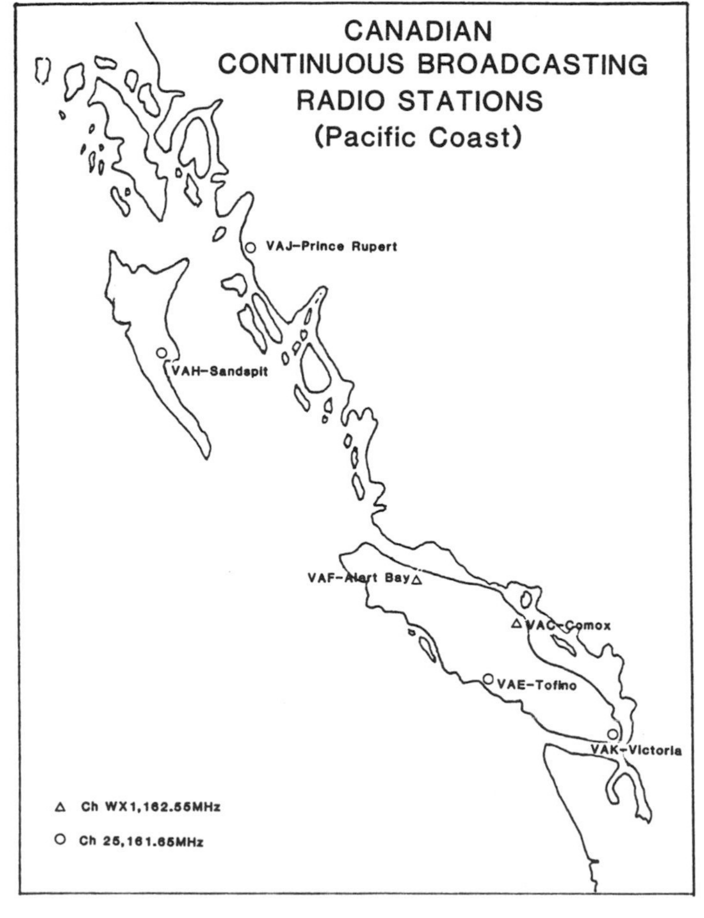

Figure 2-8. ABOVE. VHF continuous weather broadcast stations—Canada, Great Lakes. Courtesy: Worldwide Marine Weather Broadcasts—NOAA.

Figure 2-9. LEFT. VHF continuous weather broadcast stations—Canada, Pacific Coast. Courtesy: Worldwide Marine Weather Broadcasts—NOAA.

2-8) plus 162.475 MHz which is the same as the U.S. Weather 3 channel. On the Pacific coast Canada uses 162.55 MHz, which is the same as the U.S. Weather 1 channel, plus one additional channel, Channel 25 at 161.65 MHz (Fig. 2-9).

OTHER MARINE WEATHER BROADCAST SERVICES

For those who might venture south of U.S. waters to Mexico or on to the Central American Republics, no VHF-FM coverage of coastal weather in these areas is presently available. You're going to need an AM multiband receiver, and you'd best learn Morse Code.

The National Weather Service publishes a book entitled *Selected Worldwide Marine Weather Broadcasts.* This publication contains weather broadcast schedules, both U.S. and foreign, from all over the planet, covering radiotelephone, radiotelegraph (Morse code), and radiofacsimile transmissions. These schedules list broadcast times and geographic areas covered by the broadcast information, as well as station call letters, transmitting frequencies, and station locations. Those who expect to sail outside of the areas covered by VHF transmissions should consult this book to determine what radio weather information will be available to them.

As we saw in Chapter 1, the Weather Service transmits various weather maps throughout the country via its interconnected computer network. There is also a worldwide network of weather broadcast stations transmitting weather maps called radiofacsimilies (Fig. 2-10) to ships at sea. Radiofacsimilies of satellite weather photographs are also transmitted by these same stations.

A radiofacsimile receiver (Fig. 2-11) can come in an extremely compact package. The receiver illustrated measures about 15 by 16 by 4½ inches and weighs about 25 pounds. The major disadvantage is that these receivers are quite expensive. However, if you are planning on doing extensive blue water cruising, the timely receipt of weather maps and satellite photographs by this method could well be invaluable.

MARINE WEATHER SERVICES CHARTS

In addition to its book containing worldwide listings of marine weather broadcasts, the National Weather Service also publishes a series of weather services charts (Fig. 2-12). These charts show the locations of the various services around the U.S.

Of prime importance here is the location of the Weather Service VHF-FM transmitters along with their call letters and broadcast fre-

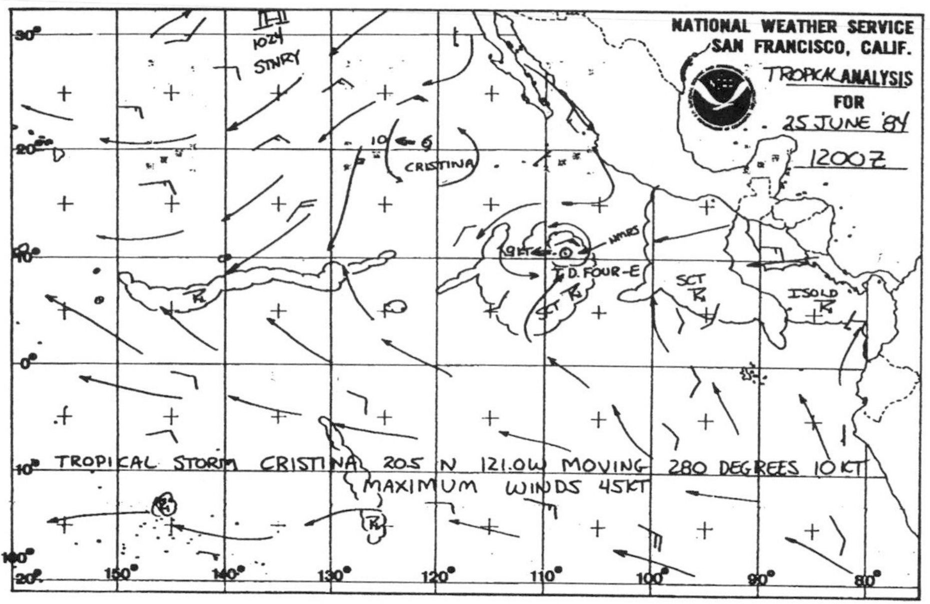

Figure 2-10. Radiofacsimile weather map as received at sea. Courtesy: Taiyo Musen Co., Ltd.

quencies. Also shown are the locations where weather warnings by flags and lights are displayed. In addition to locating the transmitters, these charts also give the locations of reporting shore stations and offshore buoys that provide the information and observations included in the VHF broadcasts.

Weather service publishes a total of fifteen of these weather charts covering the entire coastline of the continental U.S. plus Alaska, the Hawaiian Islands, and Puerto Rico and the Virgin Islands. These charts are:

MSC-1 Eastport, Me. to Montauk Pt. N.Y.
MSC-2 Montauk Pt., N.Y. to Manasquan, N.J.
MSC-3 Manasquan, N.J. to Cape Hatteras, N.C.
MSC-4 Cape Hatteras, N.C. to Savanah, Ga.
MSC-5 Savanah, Ga. to Apalachicola, Fla.
MSC-6 Apalachicola, Fla. to Morgan City, La.
MSC-7 Morgan City, La. to Brownsville, Tex.
MSC-8 Mexican Border to Pt. Conception, Calif.
MSC-9 Pt. Conception, Calif. to Pt. St. George, Calif.
MSC-10 Pt. St. George, Calif. to Canadian Border
MSC-11 Great Lakes: Michigan and Superior
MSC-12 Great Lakes: Huron, Erie, and Ontario
MSC-13 Hawaiian Waters
MSC-14 Puerto Rico and Virgin Islands
MSC-15 Alaskan Waters

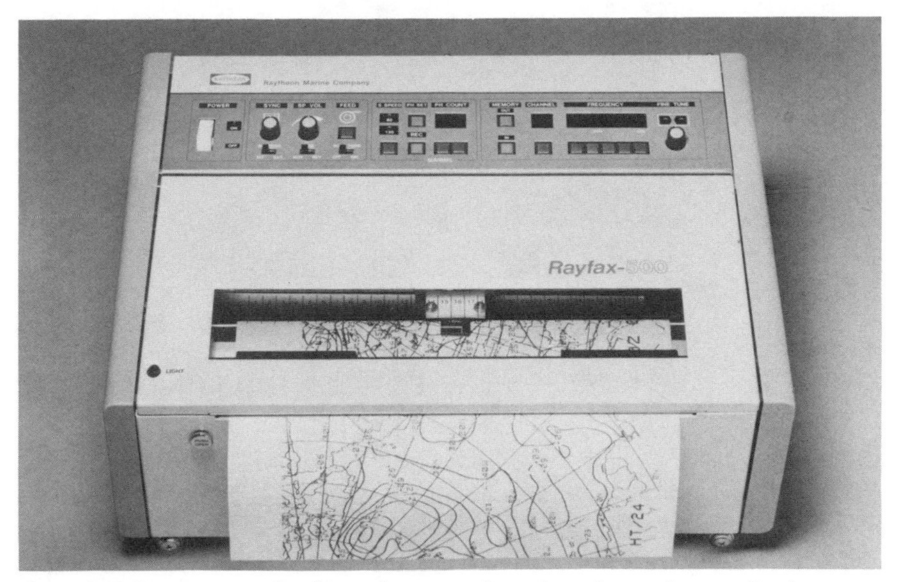

Figure 2-11. Compact weather facsimile receiver for onboard reproduction of radio transmitted weather maps. Courtesy: Raytheon Marine Co.

All of these charts are available from:

National Ocean Service
Distribution Branch (N / CG33)
Riverdale, Md. 20737

Each chart carries some general notes on wind, tide, or weather that are significant in the area covered. Also on the back of each chart is a listing of commercial radio stations in the area that normally carry marine forecasts and warnings as well as miscellaneous useful information. The MSC chart that covers your operating area should be kept aboard along with your navigational charts so as to be available for ready reference.

PILOT CHARTS

Every year, for each month of the year, the Defense Mapping Agency publishes a Pilot Chart of the North Atlantic Ocean, and a Pilot Chart of the North Pacific Ocean. There are thus twelve Pilot charts (Fig. 2-13) for each ocean. There are produced primarily for use on large ships making transoceanic passages, however, they also contain useful information for the small boatman making coastal passages.

The information on these charts is based on Weather Service records. Each chart shows wind and current directions and velocities to be anticipated at many points all over both oceans during its month

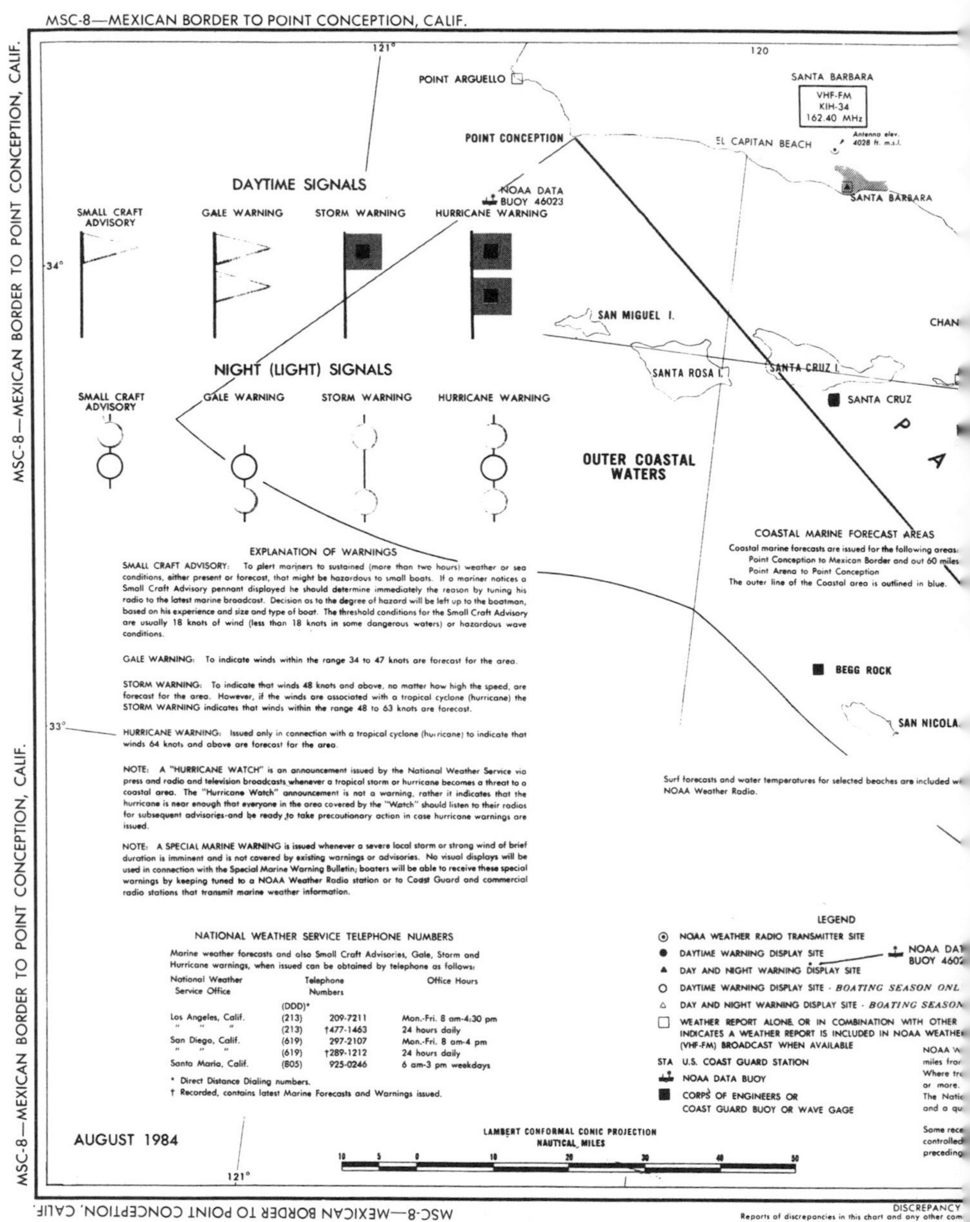

Figure 2-12. Marine weather services chart. Courtesy: National Weather Service.

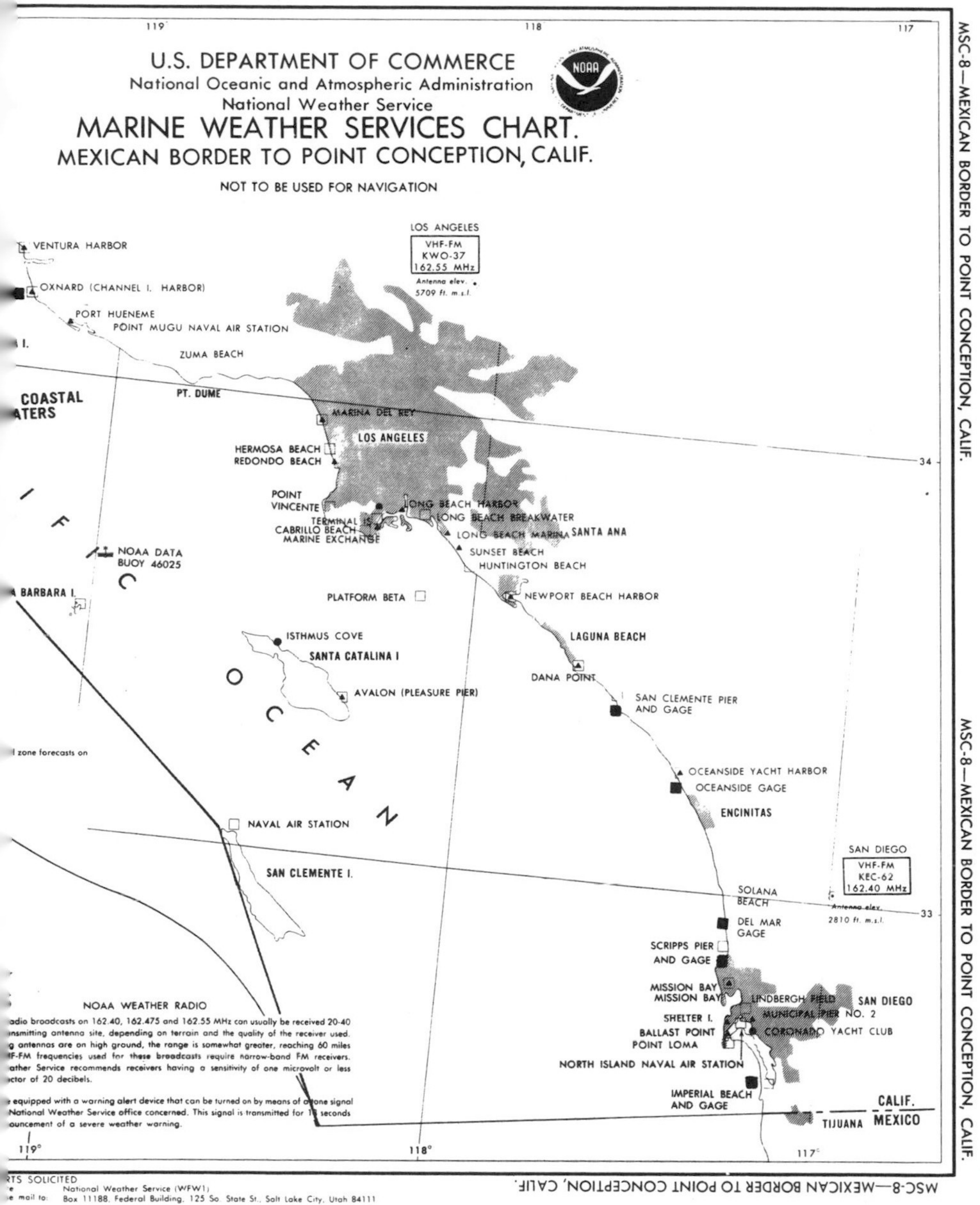
U.S. DEPARTMENT OF COMMERCE
National Oceanic and Atmospheric Administration
National Weather Service
MARINE WEATHER SERVICES CHART
MEXICAN BORDER TO POINT CONCEPTION, CALIF.
NOT TO BE USED FOR NAVIGATION
NOAA
LOS ANGELES
VHF-FM
KWO-37
162.55 MHz
Antenna elev. 5709 ft. m.s.l.
VENTURA HARBOR
OXNARD (CHANNEL I. HARBOR)
PORT HUENEME
POINT MUGU NAVAL AIR STATION
ZUMA BEACH
PT. DUME
COASTAL
MARINA DEL REY
LOS ANGELES
HERMOSA BEACH
REDONDO BEACH
POINT VINCENTE
TERMINAL IS.
CABRILLO BEACH
MARINE EXCHANGE
LONG BEACH HARBOR
LONG BEACH BREAKWATER
LONG BEACH MARINA
SANTA ANA
SUNSET BEACH
HUNTINGTON BEACH
NOAA DATA BUOY 46025
PLATFORM BETA
NEWPORT BEACH HARBOR
LAGUNA BEACH
ISTHMUS COVE
SANTA CATALINA I
AVALON (PLEASURE PIER)
DANA POINT
SAN CLEMENTE PIER AND GAGE
OCEAN
OCEANSIDE YACHT HARBOR
OCEANSIDE GAGE
ENCINITAS
NAVAL AIR STATION
SAN CLEMENTE I.
SAN DIEGO
VHF-FM
KEC-62
162.40 MHz
Antenna elev. 2810 ft. m.s.l.
SOLANA BEACH
DEL MAR GAGE
SCRIPPS PIER AND GAGE
MISSION BAY
MISSION BAY
LINDBERGH FIELD
SAN DIEGO
MUNICIPAL PIER NO. 2
SHELTER I.
BALLAST POINT
POINT LOMA
CORONADO YACHT CLUB
NORTH ISLAND NAVAL AIR STATION
IMPERIAL BEACH AND GAGE
CALIF.
MEXICO
TIJUANA
NOAA WEATHER RADIO
119
118
117
34
33
MSC-8—MEXICAN BORDER TO POINT CONCEPTION, CALIF.

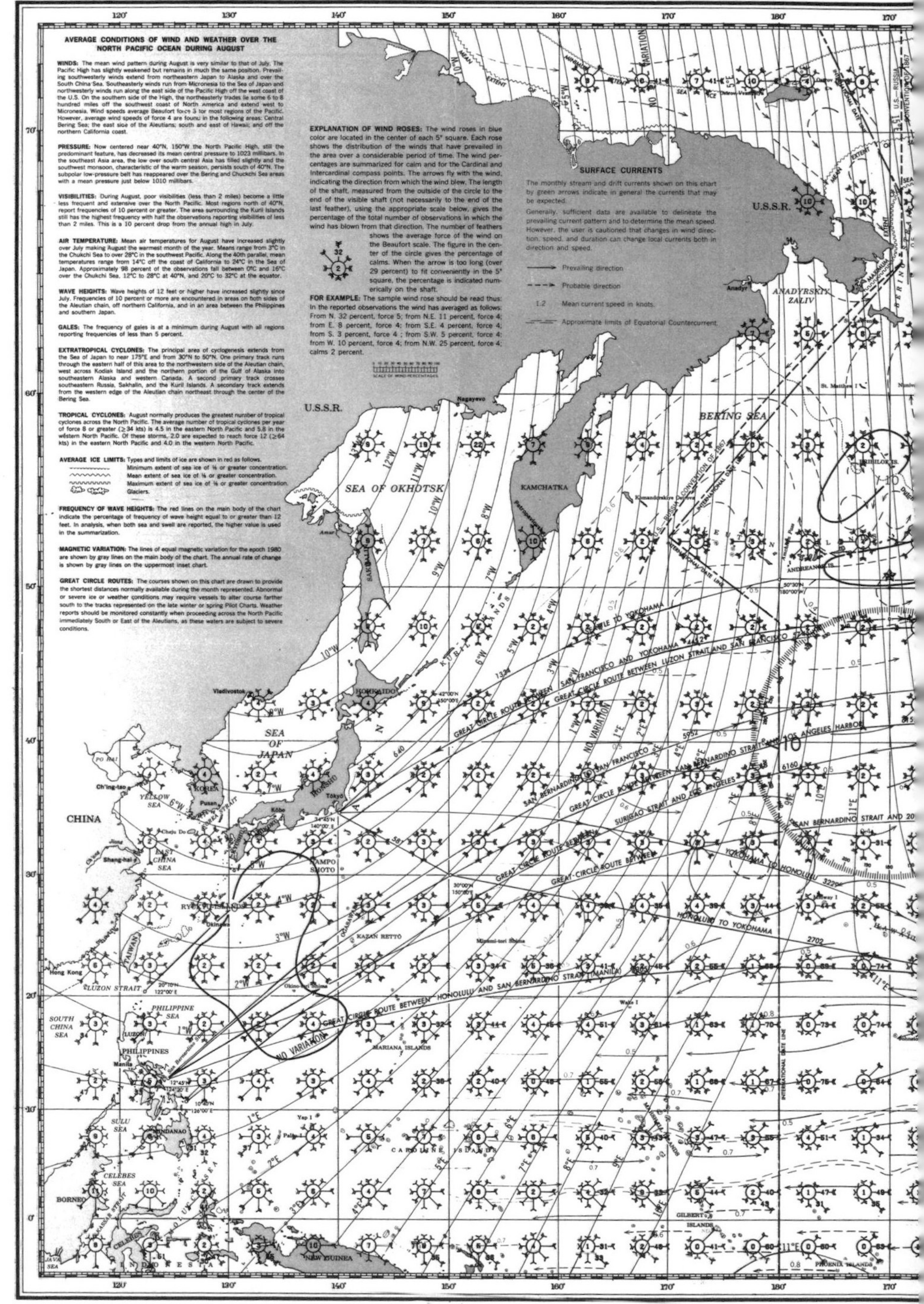

Figure 2-13. Pilot chart of the North Pacific Ocean for the month of August. Contains a wealth of information on winds, currents, wave heights, and storm frequencies to be expected at places on and around the Pacific during August. There is a chart for every month of the year.

ORTH PACIFIC OCEAN

Founded upon the researches made in the early part of the nineteenth century by Matthew Fontaine Maury, while serving as a Lieutenant in the United States Navy.

MERCATOR PROJECTION
SCALE 1:20,300,000

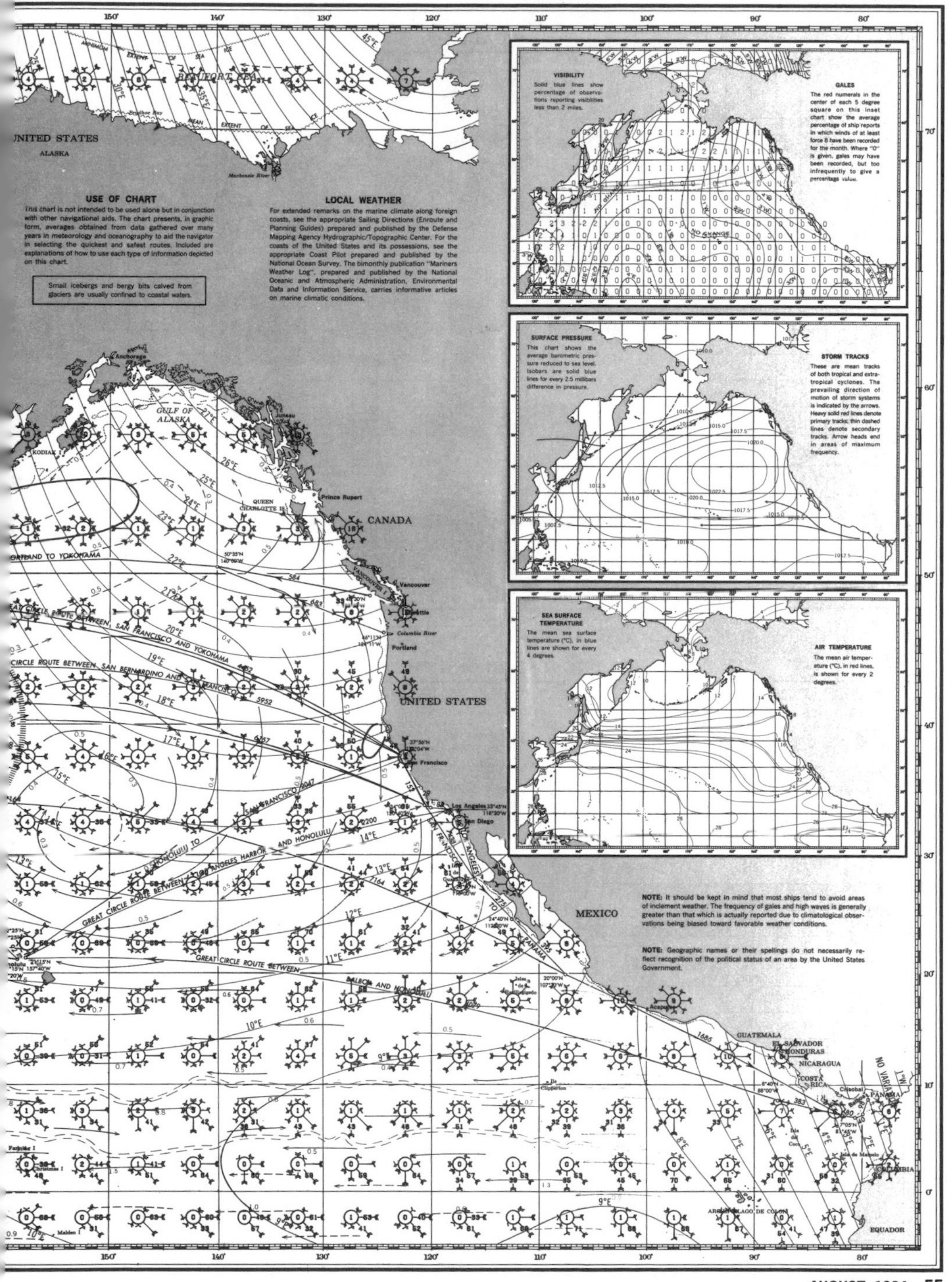

UNITED STATES
ALASKA
Mackenzie River
USE OF CHART
This chart is not intended to be used alone but in conjunction with other navigational aids. The chart presents, in graphic form, averages obtained from data gathered over many years in meteorology and oceanography to aid the navigator in selecting the quickest and safest routes. Included are explanations of how to use each type of information depicted on this chart.
Small icebergs and bergy bits calved from glaciers are usually confined to coastal waters.
LOCAL WEATHER
For extended remarks on the marine climate along foreign coasts, see the appropriate Sailing Directions (Enroute and Planning Guides) prepared and published by the Defense Mapping Agency Hydrographic/Topographic Center. For the coasts of the United States and its possessions, see the appropriate Coast Pilot prepared and published by the National Ocean Survey. The bimonthly publication "Mariners Weather Log", prepared and published by the National Oceanic and Atmospheric Administration, Environmental Data and Information Service, carries informative articles on marine climatic conditions.
VISIBILITY
Solid blue lines show percentage of observations reporting visibilities less than 2 miles.
GALES
The red numerals in the center of each 5 degree square on this inset chart show the average percentage of ship reports in which winds of at least force 8 have been recorded for the month. Where "0" is given, gales may have been recorded, but too infrequently to give a percentage value.
SURFACE PRESSURE
This chart shows the average barometric pressure reduced to sea level. Isobars are solid blue lines for every 2.5 millibars difference in pressure.
STORM TRACKS
These are mean tracks of both tropical and extratropical cyclones. The prevailing direction of motion of storm systems is indicated by the arrows. Heavy solid red lines denote primary tracks; thin dashed lines denote secondary tracks. Arrow heads end in areas of maximum frequency.
SEA SURFACE TEMPERATURE
The mean sea surface temperature (°C), in blue lines are shown for every 4 degrees.
AIR TEMPERATURE
The mean air temperature (°C), in red lines, is shown for every 2 degrees.
Anchorage
GULF OF ALASKA
KODIAK I
Juneau
Prince Rupert
QUEEN CHARLOTTE IS
CANADA
Vancouver
Seattle
Columbia River
Portland
UNITED STATES
San Francisco
Los Angeles
San Diego
MEXICO
Acapulco
GUATEMALA
EL SALVADOR
HONDURAS
NICARAGUA
COSTA RICA
PANAMA
COLOMBIA
ECUADOR
ARCHIPIELAGO DE COLON
NOTE: It should be kept in mind that most ships tend to avoid areas of inclement weather. The frequency of gales and high waves is generally greater than that which is actually reported due to climatological observations being biased toward favorable weather conditions.
NOTE: Geographic names or their spellings do not necessarily reflect recognition of the political status of an area by the United States Government.
GREAT CIRCLE ROUTE BETWEEN SAN FRANCISCO AND YOKOHAMA
GREAT CIRCLE ROUTE BETWEEN SAN BERNARDINO AND SAN FRANCISCO
GREAT CIRCLE ROUTE BETWEEN BALBOA AND HONOLULU
150°
140°
130°
120°
110°
100°
90°
80°
70°
60°
50°
40°
30°
20°
10°
0°

DMA STOCK NO.
PILOT558407

of the year. The directions and velocities shown are based on an average of what has been recorded over a great many years at the places indicated during this month.

The symbols on these charts are, at first, going to be totally unfamiliar to you, but don't panic! They are explained on the upper left-hand side of each chart. *Caution:* Since these charts are issued monthly, be sure you are not looking at the November chart in July! Also note that the information used is a statistical average developed over many years. Therefore, unlike navigational charts, you can safely consult Pilot Charts several years old because they change extremely slowly.

LAND LINE TELEPHONE

In addition to operating an elaborate network of VHF-FM radiotelephone broadcast stations on a continuous basis, the Weather Service can be reached ashore via the ordinary telephone whenever you wish. In all major cities, and a good many minor ones as well, the local Weather Service prepares tapes of the current weather forecasts, which can be heard by dialing the appropriate phone number. These tapes are updated every few hours, or as conditions change.

There used to be a page near the front of telephone directories where emergency numbers were listed for Police, Fire Department, Gas and Electric, and others along with numbers for Time and Weather. In many cases you will not find either Time or Weather at the front of the book any longer. In this case, look for "Weather Service Office" in the alphabetical listings, or if you have no luck there, find the United States Government listings. Go down to "Commerce, Department of." Under that heading will be a subhead for "National Oceanic and Atmospheric Administration." Under that you will find "Weather Service Office" with a telephone number.

The telephone weather report is similar to the one broadcast on VHF radio, but is not as complete or detailed in describing marine conditions. It too begins with a synopsis explaining current and anticipated weather system movements. It then gives a brief summary of coastal wind and wave conditions and expectations for the next 24 to 36 hours.

Since the telephone report is intended to be used by the general public, not all of whom are sailors, much of it is devoted to present and expected conditions over inland areas. The extensive forecasts for coastal and offshore waters are omitted. Also omitted are the extensive reports on present conditions at nearby coastal stations.

However, small craft and storm warnings are included.

Obviously, when unpleasant or unsafe conditions are expected over the land in a coastal area, it is safe to assume that conditions at sea along that same coast will not be delightful either. Since you probably do not have a radio at home capable of receiving the Weather Service VHF-FM broadcasts, a check of the weather as reported by telephone will normally be more than adequate to tell you whether this is likely to be a good day on the water.

ADDITIONAL SOURCES OF WEATHER INFORMATION

To a great many sailors, motorboaters, and fishermen who operate very small vessels, all this explanation of VHF and other marine radio weather broadcasts is just so much useless happy-talk. Let's say you haven't bought a VHF, or any other marine radio because of lack of space or other reasons. What, if anything, can you do to get current weather information?

There are several things you can do, but before we look at them let's make certain that you really cannot have a VHF radio aboard if you seriously want one. Several manufacturers now offer ultra compact handheld, self-powered VHF radios (see Fig. 1-13) capable of receiving all weather channels, as well as both transmitting and receiving on virtually all the VHF marine communications channels in normal use. You can even get a protective transparent waterproof case to go with these radios.

Equipped this way you could carry one of these radios on a small centerboard sailboat, or a small outboard motorboat, and capsize the boat without damaging the radio. But, let's say the budget doesn't yet permit you to get one.

COMMERCIAL RADIO

All commercial broadcast stations, AM, FM, and TV, include a weather summary along with their regular news spots. These summaries are not nearly as complete or as current as reports coming directly from the Weather Service, however, they are certainly helpful, especially when you are out in the middle of the bay away from the telephone, and the sky looks menacing.

Particularly during the summer, many commercial AM and FM radio stations run special weather broadcasts specifically for the boating public. Find out which stations in your area do this, and at what hours. You can then keep informed by using nothing more than an inexpensive AM–FM portable receiver, which you probably already carry aboard for music and entertainment.

TV WEATHER BROADCASTS

Once you are out on the bay or the lake in a small boat, TV weather broadcasts obviously aren't going to be of much use. However, the night before or even the morning before you plan to go out, it is very helpful to watch the TV weather report. This report is particularly informative because satellite pictures and weather maps are shown, enabling you to see the type and extent of the weather system moving through your area. Granted, the average TV weather reporter is far more likely to be a "personality" than a meteorologist; however, concealed in their welter of lame humor will be a pretty good coverage of the important information you need.

DAILY NEWSPAPERS

Your daily newspaper, particularly any of the big city papers, gives an excellent summary of weather conditions. Figure 2-14 is a good example of the weather summary from a city daily. Predictions are given for the immediate area of the city, then for the surrounding localities up to a couple hundred miles away.

Both a weather map for the day and a reproduction of a current satellite photograph are usually included. Between the two, you should be able to get a very clear picture of the types, sizes, and extents of the major weather systems moving across the country. As we shall see in Chapter 3, by studying these weather maps and photos for several days in a row you can trace both the direction and the speed of movement of major weather systems. By tracing these movements on maps over and over again you can develop considerable skill at predicting where weather systems will travel, and how rapidly they will get there. Then, by studying the maps for several days in advance of taking a trip, you will be able to make your own reasonably accurate weather predictions.

High and low temperatures and weather conditions in cities scattered throughout the nation are not likely to be important to you, nor are conditions in cities scattered all over the world. However, times of sunrise and sunset as well as high and low tides could be useful.

The part devoted specifically to the Marine Report is quite small, but it does briefly cover expected wind directions and velocities as well as expected sea conditions. This part is taken directly from the

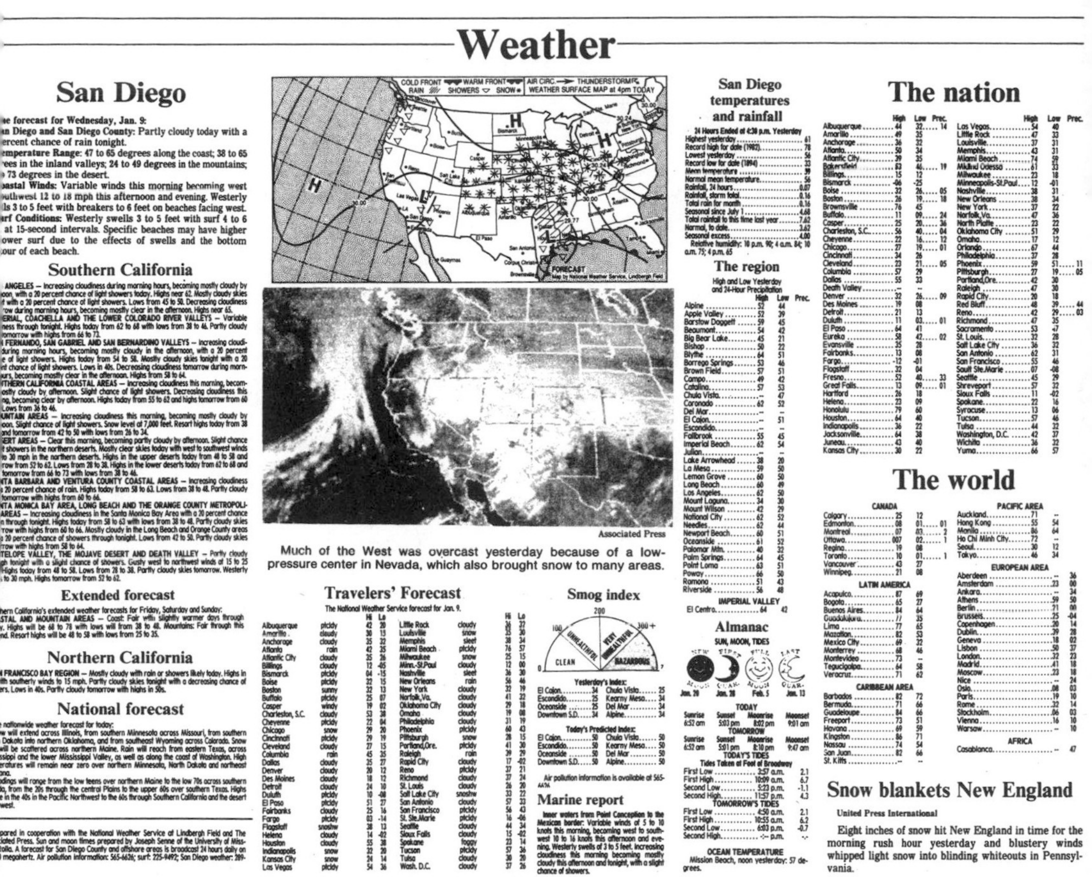

Weather

San Diego

...e forecast for Wednesday, Jan. 9:

...n Diego and San Diego County: Partly cloudy today with a ...ercent chance of rain tonight.

...emperature Range: 47 to 65 degrees along the coast; 38 to 65 ...ees in the inland valleys; 24 to 49 degrees in the mountains; ...73 degrees in the desert.

...astal Winds: Variable winds this morning becoming west ...uthwest 12 to 18 mph this afternoon and evening. Westerly ...ls 3 to 5 feet with breakers to 6 feet on beaches facing west.

...rf Conditions: Westerly swells 3 to 5 feet with surf 4 to 6 ... at 15-second intervals. Specific beaches may have higher ...ower surf due to the effects of swells and the bottom ...our of each beach.

Southern California

...ANGELES — Increasing cloudiness during morning hours, becoming mostly cloudy by ...oon, with a 20 percent chance of light showers today. Highs near 62. Mostly cloudy skies ...t with a 20 percent chance of light showers. Lows from 45 to 50. Decreasing cloudiness ...row during morning hours, becoming mostly clear in the afternoon. Highs near 65.

...ERIAL, COACHELLA AND THE LOWER COLORADO RIVER VALLEYS — Variable ...ness through tonight. Highs today from 62 to 68 with lows from 38 to 46. Partly cloudy ...omorrow with highs from 66 to 73.

...FERNANDO, SAN GABRIEL AND SAN BERNARDINO VALLEYS — Increasing cloudi... ...during morning hours, becoming mostly cloudy in the afternoon, with a 20 percent ...e of light showers. Highs today from 54 to 58. Mostly cloudy skies tonight with a 20 ...nt chance of light showers. Lows in 40s. Decreasing cloudiness tomorrow during morn... ...urs, becoming mostly clear in the afternoon. Highs from 58 to 64.

...THERN CALIFORNIA COASTAL AREAS — Increasing cloudiness this morning, becom... ...ostly cloudy by afternoon. Slight chance of light showers. Decreasing cloudiness this ...ng, becoming clear by afternoon. Highs today from 55 to 62 and highs tomorrow from 60 ...Lows from 36 to 46.

...UNTAIN AREAS — Increasing cloudiness this morning, becoming mostly cloudy by ...oon. Slight chance of light showers. Snow level at 7,000 feet. Resort highs today from 38 ...and tomorrow from 42 to 50 with lows from 26 to 34.

...SERT AREAS — Clear this morning, becoming partly cloudy by afternoon. Slight chance ...t showers in the northern deserts. Mostly clear skies today with west to southwest winds ...to 30 mph in the northern deserts. Highs in the upper deserts today from 48 to 58 and ...row from 52 to 62. Lows from 28 to 38. Highs in the lower deserts today from 62 to 68 and ...tomorrow from 66 to 73 with lows from 38 to 46.

...NTA BARBARA AND VENTURA COUNTY COASTAL AREAS — Increasing cloudiness ...a 20 percent chance of rain. Highs today from 58 to 63. Lows from 38 to 48. Partly cloudy ...tomorrow with highs from 60 to 66.

...NTA MONICA BAY AREA, LONG BEACH AND THE ORANGE COUNTY METROPOLI... ...AREAS — Increasing cloudiness in the Santa Monica Bay Area with a 20 percent chance ...n through tonight. Highs today from 58 to 63 with lows from 38 to 48. Partly cloudy skies ...rrow with highs from 60 to 66. Mostly cloudy in the Long Beach and Orange County areas ...a 20 percent chance of showers through tonight. Lows from 42 to 50. Partly cloudy skies ...rrow with highs from 58 to 64.

...TELOPE VALLEY, THE MOJAVE DESERT AND DEATH VALLEY — Partly cloudy ...h tonight with a slight chance of showers. Gusty west to northwest winds of 15 to 25 ...Highs today from 48 to 58. Lows from 28 to 38. Partly cloudy skies tomorrow. Westerly ...to 30 mph. Highs tomorrow from 52 to 62.

Extended forecast

...hern California's extended weather forecasts for Friday, Saturday and Sunday:
...STAL AND MOUNTAIN AREAS — Coast: Fair with slightly warmer days through ...y. Highs will be 68 to 78 with lows will from 38 to 48. Mountains: Fair through this ...nd. Resort highs will be 48 to 58 with lows from 25 to 35.

Northern California

...H FRANCISCO BAY REGION — Mostly cloudy with rain or showers likely today. Highs in ...th southerly winds to 15 mph. Partly cloudy skies tonight with a decreasing chance of ...ers. Lows in 40s. Partly cloudy tomorrow with highs in 50s.

National forecast

...e nationwide weather forecast for today:
...w will extend across Illinois, from southern Minnesota across Missouri, from southern ...Dakota into northern Oklahoma, and from southeast Wyoming across Colorado. Snow ...will be scattered across northern Maine. Rain will reach from eastern Texas, across ...ssippi and the lower Mississippi Valley, as well as along the coast of Washington. High ...eratures will remain near zero over northern Minnesota, North Dakota and northeast ...ana.
...dings will range from the low teens over northern Maine to the low 70s across southern ...da, from the 20s through the central Plains to the upper 60s over southern Texas. Highs ...e in the 40s in the Pacific Northwest to the 60s through Southern California and the desert ...west.

...pared in cooperation with the National Weather Service at Lindbergh Field and The ...ciated Press. Sun and moon times prepared by Joseph Senne of the University of Miss... ...ouri-Rolla. A forecast for San Diego County and offshore areas is broadcast 24 hours daily on ...0 megahertz. Air pollution information: 565-6626; surf: 225-9492; San Diego weather: 289-

Associated Press

Much of the West was overcast yesterday because of a low-pressure center in Nevada, which also brought snow to many areas.

Travelers' Forecast

The National Weather Service forecast for Jan. 9.

City		Hi	Lo
Albuquerque	ptcldy	42	20
Amarillo	cloudy	30	15
Anchorage	cloudy	35	32
Atlanta	rain	42	35
Atlantic City	cloudy	35	26
Billings	cloudy	12	-05
Bismarck	ptcldy	04	-15
Boise	ptcldy	32	15
Boston	sunny	32	13
Buffalo	ptcldy	25	07
Casper	windy	19	02
Charleston, S.C.	cloudy	53	38
Cheyenne	ptcldy	22	04
Chicago	snow	29	20
Cincinnati	ptcldy	29	19
Cleveland	cloudy	27	15
Columbia	rain	45	35
Dallas	cloudy	35	27
Denver	cloudy	20	12
Des Moines	cloudy	18	12
Detroit	cloudy	24	10
Duluth	ptcldy	10	-08
El Paso	ptcldy	51	27
Fairbanks	cloudy	25	16
Fargo	ptcldy	03	-14
Flagstaff	snoshw	38	13
Helena	cloudy	14	-02
Houston	cloudy	55	38
Indianapolis	snow	32	20
Kansas City	snow	24	14
Las Vegas	ptcldy	54	36

City		Hi	Lo
Little Rock	cloudy	36	27
Louisville	snow	35	30
Memphis	sleet	38	34
Miami Beach	ptcldy	76	57
Milwaukee	snow	25	15
Minn.-St.Paul	cloudy	12	00
Nashville	sleet	36	30
New Orleans	rain	56	46
New York	cloudy	32	19
Norfolk, Va.	cloudy	41	32
Oklahoma City	cloudy	29	18
Omaha	cloudy	19	08
Philadelphia	cloudy	31	19
Phoenix	ptcldy	60	43
Pittsburgh	snow	28	15
Portland, Ore.	ptcldy	41	30
Raleigh	rain	39	29
Rapid City	cloudy	17	-02
Reno	ptcldy	37	21
Richmond	cloudy	37	24
St. Louis	cloudy	26	20
Salt Lake City	snoshw	33	22
San Antonio	cloudy	57	33
San Francisco	ptcldy	56	43
St. Ste.Marie	ptcldy	16	-06
Seattle	cloudy	44	34
Sioux Falls	cloudy	15	-02
Spokane	foggy	23	14
Tucson	ptcldy	57	36
Tulsa	cloudy	30	20
Wash. D.C.	cloudy	37	26

Smog index

Yesterday's Index:

El Cajon	34	Chula Vista	25
Escondido	25	Kearny Mesa	34
Oceanside	25	Del Mar	34
Downtown S.D.	34	Alpine	34

Today's Predicted Index:

El Cajon	50	Chula Vista	50
Escondido	50	Kearny Mesa	50
Oceanside	50	Del Mar	50
Downtown S.D.	50	Alpine	50

Air pollution information is available at 565-AA?A

Marine report

Inner waters from Point Conception to the Mexican border: Variable winds of 5 to 10 knots this morning, becoming west to southwest 10 to 16 knots this afternoon and evening. Westerly swells of 3 to 5 feet. Increasing cloudiness this morning becoming mostly cloudy this afternoon and tonight, with a slight chance of showers.

San Diego temperatures and rainfall

24 Hours Ended at 4:30 p.m. Yesterday

Highest yesterday	61
Record high for date (1982)	78
Lowest yesterday	56
Record low for date (1894)	33
Mean temperature	59
Normal mean temperature	56
Rainfall, 24 hours	0.07
Rainfall, storm total	0.16
Total rain for month	0.16
Seasonal since July 1	4.68
Total rainfall to this time last year	7.62
Normal, to date	3.62
Seasonal excess	4.00

Relative humidity: 10 p.m. 90; 4 a.m. 84; 10 a.m. 75; 4 p.m. 65

The region

High and Low Yesterday and 24-Hour Precipitation

	High	Low	Prec.
Alpine	52	44	
Apple Valley	52	39	
Barstow Doggett	59	45	
Beaumont	54	42	
Big Bear Lake	45	21	
Bishop	50	22	
Blythe	64	51	
Borrego Springs	53	46	
Brown Field	57	51	
Campo	49	42	
Catalina	57	53	
Chula Vista	--	47	
Coronado	62	52	
Del Mar	--	--	
El Cajon	--	51	
Escondido	--	--	
Fallbrook	55	45	
Imperial Beach	62	54	
Julian	--	--	
Lake Arrowhead	38	20	
La Mesa	59	50	
Lemon Grove	60	50	
Long Beach	60	49	
Los Angeles	62	50	
Mount Laguna	34	30	
Mount Wilson	42	29	
National City	62	52	
Needles	62	44	
Newport Beach	60	51	
Oceanside	61	52	
Palomar Mtn.	40	32	
Palm Springs	64	45	
Point Loma	63	51	
Poway	66	50	
Ramona	51	43	
Riverside	56	48	

IMPERIAL VALLEY

	High	Low
El Centro	64	42

Almanac

SUN, MOON, TIDES

Jan. 20	Jan. 28	Feb. 5	Jan. 13

TODAY

Sunrise	Sunset	Moonrise	Moonset
6:52 am	5:03 pm	8:02 pm	9:01 am

TOMORROW

Sunrise	Sunset	Moonrise	Moonset
6:52 am	5:01 pm	8:10 pm	9:47 am

TODAY'S TIDES

Tides Taken at Foot of Broadway

First Low	3:57 a.m.	2.1
First High	10:09 a.m.	6.7
Second Low	5:23 p.m.	-1.1
Second High	11:57 p.m.	4.3

TOMORROW'S TIDES

First Low	4:50 a.m.	2.1
First High	10:55 a.m.	6.2
Second Low	6:03 p.m.	-0.7
Second High	-:-- p.m.	-.-

OCEAN TEMPERATURE

Mission Beach, noon yesterday: 57 degrees.

The nation

	High	Low	Prec.
Albuquerque	44	32	14
Amarillo	49	35	
Anchorage	36	32	
Atlanta	50	34	
Atlantic City	39	35	
Bakersfield	83	46	19
Billings	15	12	
Bismarck	-06	-25	
Boise	32	26	05
Boston	26	19	18
Brownsville	76	45	
Buffalo	11	09	24
Casper	25	20	36
Charleston, S.C.	56	40	04
Cheyenne	22	16	12
Chicago	27	19	01
Cincinnati	34	26	
Cleveland	23	21	05
Columbia	57	29	
Dallas	55	33	
Death Valley	--	--	
Denver	32	26	09
Des Moines	19	08	
Detroit	21	13	
Duluth	11	03	01
El Paso	64	41	
Eureka	58	42	02
Evansville	35	28	
Fairbanks	13	08	
Fargo	-12	-01	
Flagstaff	32	04	
Fresno	52	40	33
Great Falls	13	09	01
Hartford	26	18	
Helena	23	09	
Honolulu	79	60	
Houston	64	40	
Indianapolis	36	22	
Jacksonville	64	38	
Juneau	43	40	
Kansas City	30	22	

	High	Low	Prec.
Las Vegas	54	40	
Little Rock	47	33	
Louisville	37	31	
Memphis	43	31	
Miami Beach	74	59	
Midland Odessa	61	33	
Milwaukee	23	18	
Minneapolis-St.Paul	12	-01	
Nashville	38	31	
New Orleans	38	34	
New York	37	22	
Norfolk, Va.	47	36	
North Platte	23	22	
Oklahoma City	51	29	
Omaha	17	12	
Orlando	67	44	
Philadelphia	37	28	
Phoenix	59	51	11
Pittsburgh	27	19	05
Portland, Ore.	42	30	
Raleigh	47	30	
Rapid City	20	18	
Red Bluff	48	39	44
Reno	42	29	03
Richmond	47	35	
Sacramento	53	47	
St. Louis	32	28	
Salt Lake City	36	32	
San Antonio	62	31	
San Francisco	55	46	
Sault Ste.Marie	07	-08	
Seattle	45	29	
Shreveport	57	32	
Sioux Falls	11	-02	
Spokane	22	16	
Syracuse	13	06	
Tucson	57	46	
Tulsa	44	32	
Washington, D.C.	42	37	
Wichita	36	32	
Yuma	66	57	

The world

CANADA

Calgary	25	12	
Edmonton	08	01	01
Montreal	07	03	2
Ottawa	007	02	1
Regina	19	08	
Toronto	10	01	1
Vancouver	43	27	
Winnipeg	21	08	

LATIN AMERICA

Acapulco	87	69
Bogota	65	27
Buenos Aires	84	64
Guadalajara	77	35
Lima	77	65
Mazatlan	82	53
Mexico City	69	32
Monterrey	68	46
Montevideo	73	--
Tegucigalpa	64	58
Veracruz	71	62

CARIBBEAN AREA

Barbados	82	72
Bermuda	71	66
Guadeloupe	84	66
Freeport	73	51
Havana	69	59
Kingston	86	71
Nassau	74	54
San Juan	82	66
St. Kitts	--	--

PACIFIC AREA

Auckland	71	--
Hong Kong	55	54
Manila	86	64
Ho Chi Minh City	72	--
Seoul	30	12
Tokyo	46	34

EUROPEAN AREA

Aberdeen	--	36
Amsterdam	23	00
Ankara	--	34
Athens	59	50
Berlin	71	00
Brussels	25	-04
Copenhagen	20	14
Dublin	39	28
Geneva	18	02
Lisbon	50	37
London	32	23
Madrid	41	18
Moscow	23	18
Nice	--	24
Oslo	08	03
Paris	19	10
Rome	32	14
Stockholm	06	03
Vienna	16	10
Warsaw	--	10

AFRICA

Casablanca	--	47

Snow blankets New England

United Press International

Eight inches of snow hit New England in time for the morning rush hour yesterday and blustery winds whipped light snow into blinding whiteouts in Pennsylvania.

Figure 2-14. Typical weather reports in large city newspapers.

Weather Service coastal marine prediction that is broadcast on VHF radio as discussed earlier.

The biggest disadvantage of newspaper weather reports is that they are based on comparatively old information. Radio and TV reports give you more recent information than newspapers. Reports direct from Weather Service via telephone or VHF radio are the most recent you can normally get. The huge advantage found in newspaper reports is the weather map and satellite photo. Like their TV counterparts, these show you where weather systems are, and how they are moving, but in addition, they'll politely stand still for a while so you can study them in depth.

PART II

WEATHER AND THE SMALL-BOAT SAILOR

3

ONBOARD OBSERVATIONS SUPPLEMENTING PROFESSIONAL FORECASTS

The most highly localized professional Weather Bureau forecasts you can obtain in either broadcast or printed form cover areas that are, in relation to a single boat, quite large. Within such areas many really localized variations can, and do, occur. Hills, valleys, lakes, rivers, bays, and other natural features may cause marked differences in wind direction or velocity, as well as air temperature, humidity, and stability.

In addition, there is always the horrid possibility that a major weather disturbance, although long tracked accurately by Weather Service, will fail to move as expected for any of several reasons. In Part III, dealing with the mechanics of weather, we shall see that the prediction of the future actions of weather systems is based largely on their past movements. Generally a weather system will move and act for the next few days in a manner consistent with that of the last few days.

The professional forecaster, in projecting upcoming weather must also factor in expected interactions between nearby systems. At times, for various reasons, they do not interact as expected. In spite of the elaborate information gathering apparatus described in Chapter 1, the forecaster is still working with information that is unavoidably incomplete. From time to time, influences unknown to the forecaster produce unforeseen results.

Thus, even with the best available forecasts in our hands, you and I, from time to time, encounter the unexpected—unexpected that is if *we ourselves fail to keep our own onboard weather watch*. This, in turn, means training ourselves as weather observers in the manner traditional among seafaring people.

WEATHER INSTRUMENTS

To maintain an adequate weather watch, a little elementary instrumentation is necessary. Depending on the size of your boat, and the distance and duration of the trips you take, there are several weather instruments you should consider having aboard.

Prices for these instruments vary widely. Highly accurate ones are, of course, quite expensive. However, extremely exact readings of temperature or barometric pressure, for example, are not critical for our purposes. We shall see as we proceed, that change is what really concerns us. The directions and rates of changes over time are more significant than exact readings at any particular time. A thermometer or a barometer that reads consistently too high or too low, will still give us a clear indication of the direction and rate of temperature or pressure changes.

BAROMETER

The barometer invented by Torricelli was the *mercury barometer*. The mercury barometer works perfectly as long as it is mounted in a fixed position where it will not be disturbed. On a boat, obviously, it is going to be disturbed—at times considerably!

However, another instrument for measuring atmospheric pressure has been developed. It is the *aneroid barometer* (Fig. 3-1). The aneroid barometer is not as accurate as the mercury type, but it is comparatively inexpensive, and sufficiently rugged to operate properly at sea. Its heart is a little corrugated metal box from which most of the air is removed. This empty container then contracts and expands, reflecting changes in atmospheric pressure. Through a lever and gear mechanism, the small movements of the box are amplified and transferred to a pivoting needle that indicates the pressure reading on a curved scale. An adjustment screw in the back of your aneroid barometer allows you to set it to agree with the reading from the Weather Service.

Barometer readings decrease with altitude because atmospheric pressure decreases with altitude. Consequently, if you live up on a hill a couple hundred feet high, do not set your boat barometer at

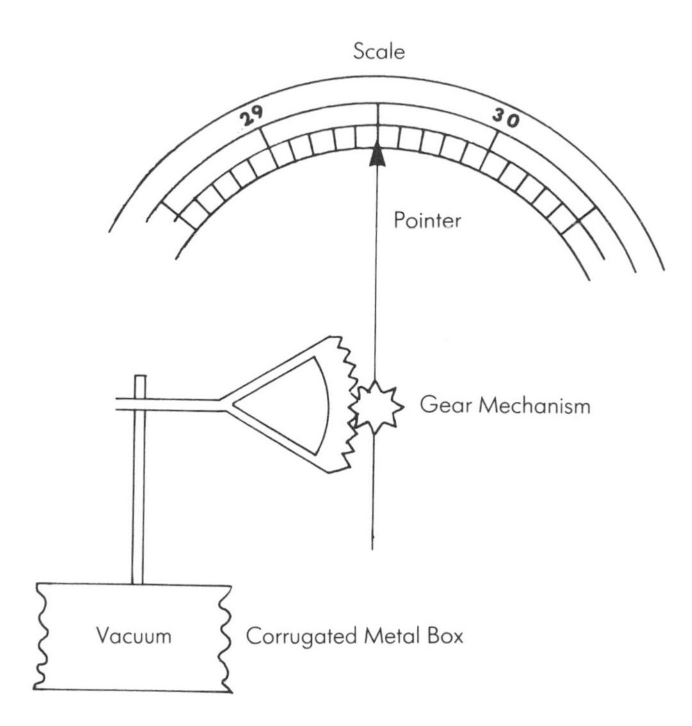

Figure 3-1. Working parts of an aneroid barometer.

home to agree with the weather broadcast. By the time you get down to the boat at sea level the barometer will have risen even though sea level pressure has remained unchanged. The aircraft instrument used to measure altitude, the altimeter, is at heart simply an aneroid barometer. It measures altitude of the aircraft by the reduction in atmospheric pressure.

READING THE BAROMETER

The barometer you will have on board will probably be calibrated in inches harking back to the original mercury barometers. A good many weather broadcasts, particularly those emanating directly from Weather Service, refer to atmospheric pressure in terms of *millibars*. Your barometer may indicate both inches and millibars.

The standard atmospheric pressure at sea level is 1013 millibars, or 29.91 inches (of mercury). One millibar equals .02953 inches or in reverse 1 inch equals 33.864 millibars. (On your handy pocket calculator divide 1013 by 33.864. The answer is 29.913773.) This standard pressure is really a sort of mid-point around which the sea level pressure fluctuates. Actually, it seldom hits exactly 29.91 inches or stays there for any length of time.

Looking again at a typical aneroid barometer scale (Fig. 3-2), the

Figure 3-2. Marine type aneroid barometer. Courtesy: Weems & Plath.

full inches—28, 29, 30, and 31—are clearly numbered. Usually the ½ inch points are marked with a "5" meaning .5 inches. On better barometers each tenth of an inch (.1) is marked by numbers running from 1 through 9. Each tenth space is then further split into 5 subdivisions each standing for two hundredths of an inch (.02). Barometer readings can thus be conveniently made to hundredths of an inch.

When the barometer is falling or rising rapidly, it means that a change in the weather, for the worse, or the better, is likely to occur quickly. When the barometer is changing slowly, the weather will change slowly as well. Fine! But what exactly is the definition of rising or falling rapidly or slowly? The U.S. Weather Service defines a rapidly rising or falling barometer as a rate of 6 hundredths of an inch (.06) per hour or faster. Slowly rising or falling thus becomes 5 hundredths (.05) or less per hour.

ATMOSPHERIC PRESSURE VARIATIONS

The atmospheric pressure as shown on the barometer moves up and down along with changes in the weather. Storms are normally centers of low pressure. Consequently, a falling barometer—indicating a lowering of atmospheric pressure —is a sign that normally precedes advancing storms. As may be seen in Figure 11-5, an advancing cyclonic storm center, when shown on a weather map, is surrounded by more or less concentric rings called *isobars,* of decreasing pressure. An isobar is simply a line on a weather map connecting places that had the same atmospheric pressure at the time the map was drawn.

Note: a falling barometer does not always prove that a storm is coming. However, the reverse is true—an approaching storm will invariably produce a falling barometer. The important key to the

Figure 3-3. Marine dial type thermometer, combined in this instrument with hygrometer (measures humidity). Thermometer is left-hand scale, reading centigrade on top (10, 20, 30, 40) and Fahrenheit below (40, 60, 80, 100). Hygrometer scale is on right side reading 20 to 100. Courtesy: Weems & Plath.

severity of the advancing storm is how far the barometer falls, and how *rapidly.*

In the same way that a falling barometer accompanies an oncoming storm, a rising barometer normally accompanies approaching fair weather. How far it rises, and how fast, is the key indicator of how quickly the weather will improve.

Atmospheric pressure changes associated with changing weather should not be confused with the normal daily periodic pressure variations. Each day the air pressure moves up and down twice. The peaks occur at about 10 AM and 10 PM with the lows about halfway between at 4 AM and 4 PM.

THERMOMETER

Air temperature along with atmospheric pressure is a major indicator of weather changes. As we shall see in Chapter 10, major weather disturbances are produced by the clashing of air masses having distinctly different characters. Temperature, in addition to pressure, is a primary characteristic in defining air masses.

A thermometer consisting of a mercury column enclosed in a glass tube is, like the mercury barometer, rather delicate for marine use. But more important it is not easy to read in poor light on a bouncing boat. Thus, your marine thermometer (Fig. 3-3) will have a round dial and pointer similar to your barometer, making it easy to read under the often less than ideal conditions at sea.

TEMPERATURE VARIATIONS

Air above warm water or warm land will be heated by contact with any surface warmer than it is. This heating will become apparent as

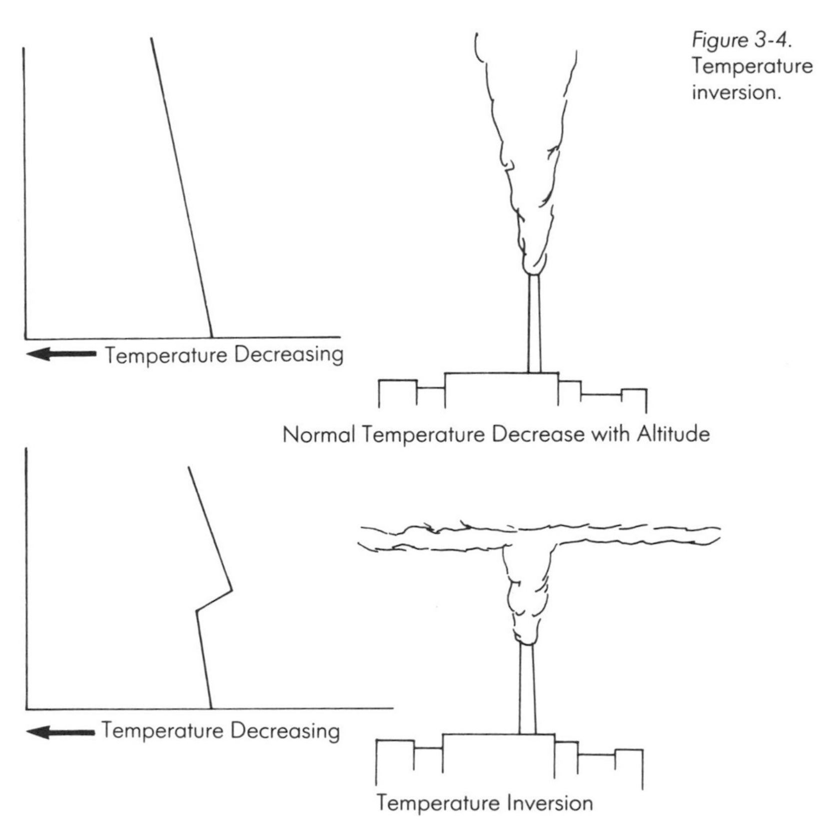

Figure 3-4. Temperature inversion.

an increase in the thermometer reading. Similarly, air above a cold surface will be cooled, reducing the thermometer reading. Thus sea level air temperatures vary *horizontally* in response to the temperatures of the underlying surfaces.

From sea level we are not able to read *vertical* variations in air temperature directly, but we can see evidence of such variations in many ways. In the *troposphere,* or lower atmosphere, air temperature normally tends to decrease with altitude (see Fig. 8-1) at a fairly steady rate. Various factors continually disrupt this "normal" trend.

Warm air wants to expand and rise, as demonstrated by smoke rising and expanding through still air, often visible as an inverted cone above a chimney (Fig. 3-4). Air warmed by contact with earth that has been heated by the sun rises and expands in the same way. This movement does not become visible until the rising air has cooled enough to start condensing the water vapor in it into puffy clouds of a type called *cumulus* (Fig. 3-5).

From time to time, horizontal air movements carry warm air, which tends to rise, up above cooler air which is heavier, denser, and hence will sink. This forms an *inversion.* In an inversion, the air temperature

Figure 3-5. Cumulus clouds form in rising air over sun-warmed land.

starts its normal decrease with altitude from sea level upward through the layer of cool air until the layer of warm air above it is reached (Fig. 3-6). At that point the air temperature rises with altitude through the warm layer, and then begins to drop again.

When you see smoke rising in the usual manner but reaching a level where it ceases to rise and simply spreads out, it has probably reached an inversion. When you look off in the distance, and see a layer of low hazy air with clear air above it, the division probably marks an inversion.

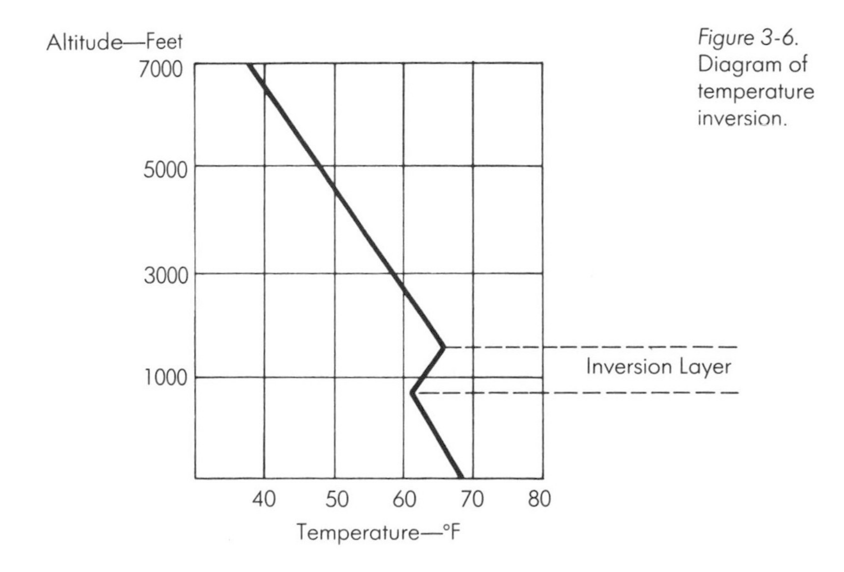

Figure 3-6. Diagram of temperature inversion.

On land, very large surface air temperature variations between day and night are common. At sea, these daily, or diurnal, variations are greatly reduced. This results from the fact that sunlight raises the surface temperature of earth very rapidly. By contrast, the surface temperature of water changes very slowly (see Chapter 8).

TEMPERATURE AND PRESSURE

In the study of weather, air temperature and air pressure are inseparably connected. As air warms, it expands, decreasing atmospheric pressure. Cold air contracts, increasing atmospheric pressure. Changes in temperature thus produce changes in pressure, or alternately, when changes in pressure are seen, expect changes in temperature as well.

It may be rainy and stormy right now, but if the temperature starts to drop sharply along with a rising barometer, be of good cheer—the weather will soon improve. Cool, high pressure, clear air is on the way.

Conversely, a lowering barometer accompanied by increasing temperatures often indicate poor weather coming (see Chapters 10 and 11).

WIND SPEED INDICATOR—ANEMOMETER

Along with atmospheric pressure and temperature, wind direction and velocity are important weather signs. Several quite elaborate wind indicating systems are available (Fig. 3-7) for small boat use. These systems are particularly useful on racing sailboats. The *apparent wind* speed is read from the rate at which the wind turns a spinner mounted at the mast head. The apparent wind direction is transmitted from a flyer also at the masthead.

For those who have no need for such an elaborate wind indicating system a small, handheld wind indicator is available (see Fig. 1-14). This device requires no installation or power supply, and virtually no storage space. It is by no means a "high-tech" instrument, but it will indicate approximate wind speeds. Also, by swinging it from side to side to find maximum readings, you can find the apparent wind direction.

The term *apparent wind* is familiar to most sailboat sailors, but others may need some explanation. When your boat is at anchor or at the dock, the wind direction and velocity you feel is the *true* wind (Fig. 3-8a). However, once the boat is under way, this is no longer so.

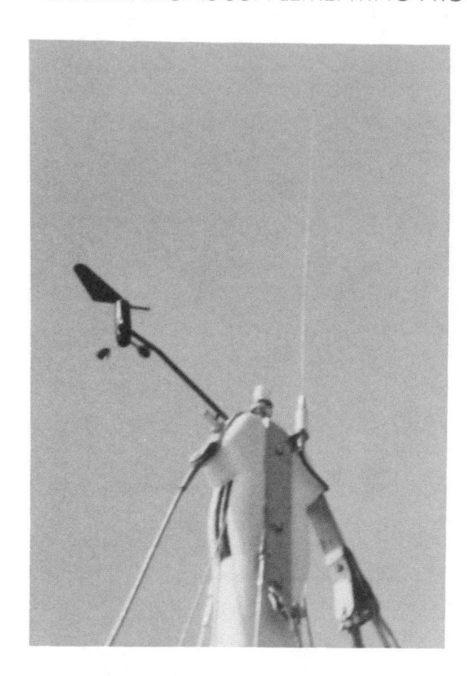

Figure 3-7. Wind speed and direction indicator.
LEFT. Masthead sending unit.
BELOW LEFT. Cockpit wind direction dial (relative direction).
BELOW RIGHT. Cockpit wind speed dial.

When the boat is heading directly into the true wind, both true and apparent wind *directions* remain the same (Fig. 3-8b), but the velocity of the apparent wind is the sum of the true wind speed plus the boat speed. If the boat is heading directly away from the true wind, both true and apparent wind directions are still unchanged, but apparent wind *speed* is now the true speed minus boat speed (Fig. 3-8c). When the boat sails at an angle to the true wind, its speed through the air becomes a vector that moves the apparent wind *forward* relative to the boat, and increases the speed of the apparent wind at the same time (Fig. 3-8d).

The boatman should be well aware of the difference between true

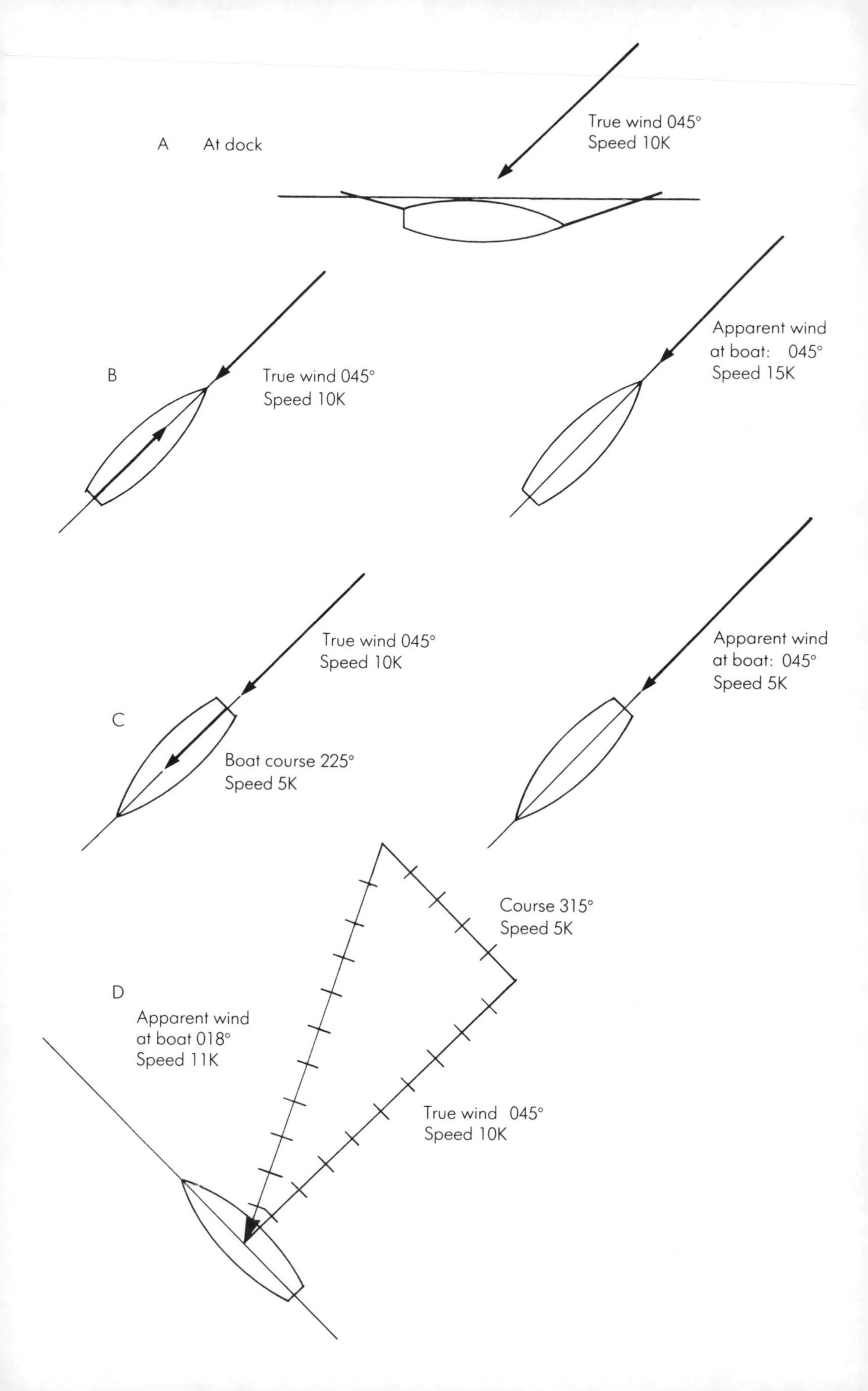
A
At dock
True wind 045°
Speed 10K
B
True wind 045°
Speed 10K
Apparent wind
at boat: 045°
Speed 15K
C
True wind 045°
Speed 10K
Boat course 225°
Speed 5K
Apparent wind
at boat: 045°
Speed 5K
D
Course 315°
Speed 5K
Apparent wind
at boat 018°
Speed 11K
True wind 045°
Speed 10K

and apparent wind because while under way no estimates and no instrument readings of either wind speed or direction will be true! Wind speeds will usually be read or estimated too high (except for a following wind), and the apparent wind direction will be moved forward toward the bow of the boat.

For a familiar example of the effect of your movement on the apparent direction of the wind, remember what happens when you drive a car. Let's say you are stopped at a gas station. The wind is coming from your left at about 10 mph. When you move out onto the highway and are driving along at 55 mph the wind you feel is now directly in front of you, right? The apparent wind has moved forward and increased immensely.

NORMAL FAIR WEATHER WIND PATTERNS

Wherever you sail in the U.S., whether it's around Cape Cod, Chesapeake Bay, San Francisco Bay, or Mobile Bay, the prevailing fair weather winds will be from the western quadrant ranging from northwest down to southwest. Average fair weather wind velocities vary widely from place to place and from month to month. Your personal experience and that of other sailors in your vicinity should guide you as to what conditions are normal for the area and season of the year.

Particularly in the summer, there are many sections of our coasts where the normal prevailing fair weather westerlies are very light. A local daily, alternating *land breeze / sea breeze* sequence may then become dominant.

What's happening is very simple. The morning sun heats the land making it much warmer than the adjoining water. The warmed ground transfers heat to the air above it (Fig. 3-9) which expands. It's pressure decreases forming a localized *thermal low*. This air becomes lighter and rises. As it rises, it is replaced by cooler air that moves in from sea, producing the daytime *sea breeze*.

At night the process is reversed. The land cools rapidly after sundown, cooling a low layer of air directly above it. The sea transfers heat to the air above it which rises to be replaced by cool air from the land, producing the familiar night *land breeze*. The daytime sea breeze is usually considerably stronger than the night land breeze.

Depending on the topography of the land and the character of the vegetation, the land breeze / sea breeze regime may penetrate considerably further inland than Figure 3-9 indicates. On occasion it may work as far as 50 miles or more inland.

Figure 3-8. OPPOSITE.

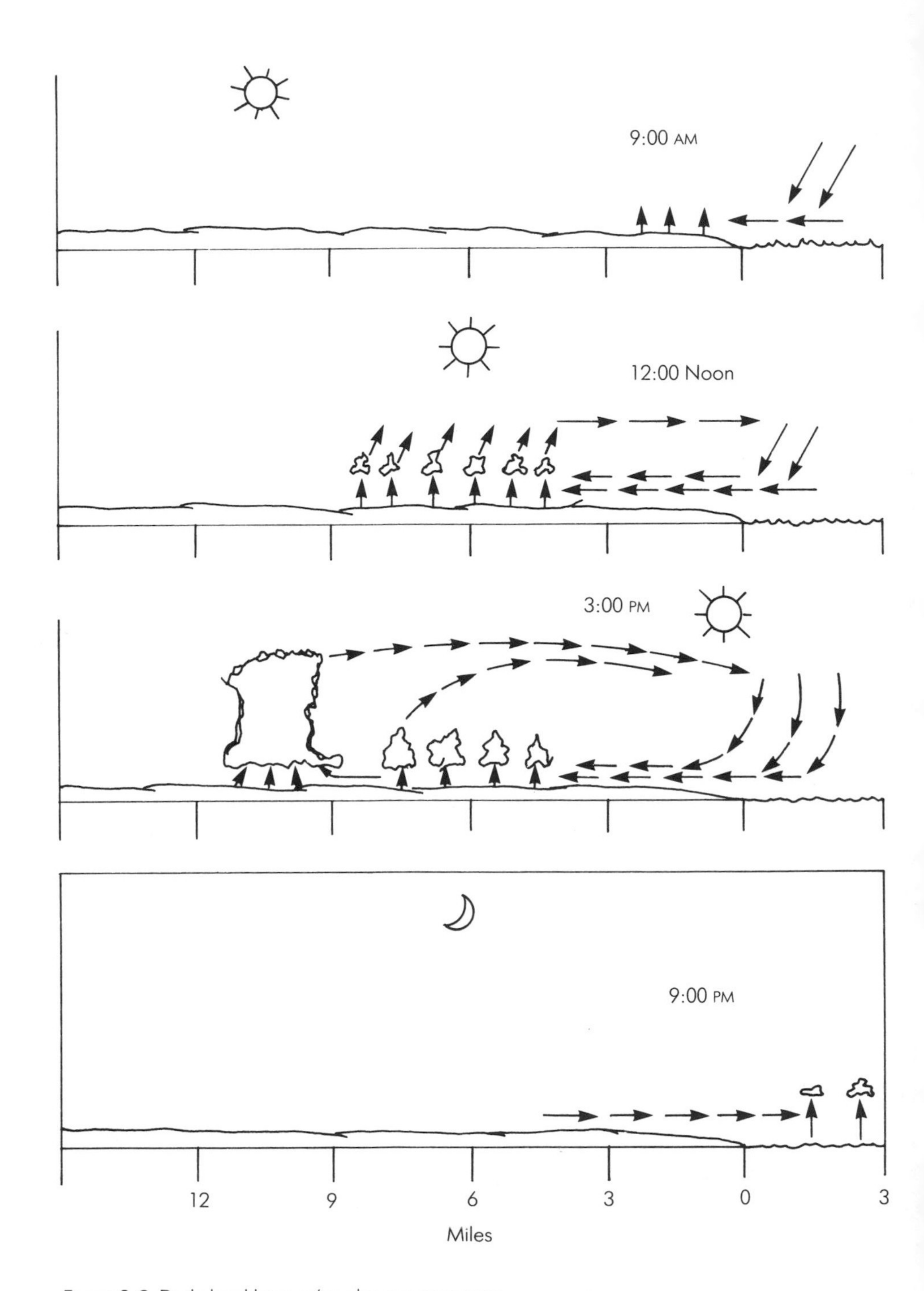

Figure 3-9. Daily land breeze/sea breeze sequence.

WINDS AND WEATHER CHANGES

Changes in atmospheric pressure and temperature are reliable signs of impending weather changes. These air pressure and temperature changes also produce alterations in wind direction and velocity. Such shifts in the winds are further indicators of future weather.

In general, when the wind *backs,* poorer weather is coming. A backing wind is one that shifts in direction counterclockwise. Let's say in the morning the wind is from the southwest. Then during the day it shifts to the south, then southeast, and finally east (Fig. 3-10). This is a typical "backing wind" sequence. Wind velocity is probably increasing and is very likely to be accompanied by a falling barometer, thickening and lower clouds, and slowly rising temperature, all of which precede poor weather.

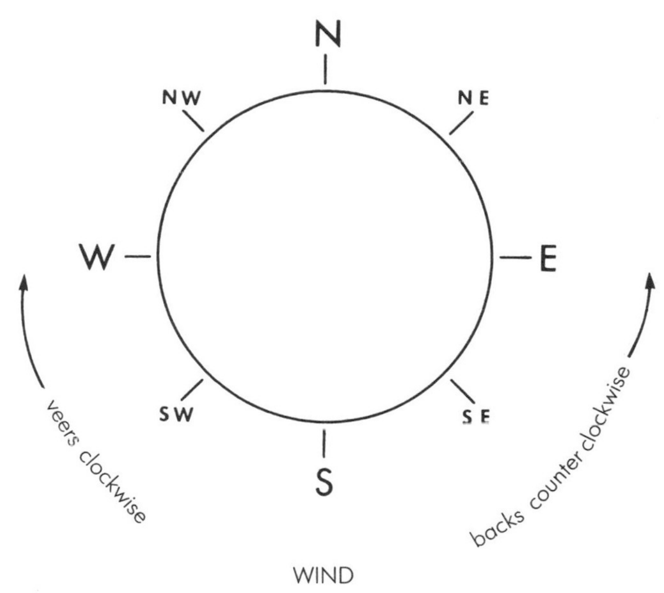

Figure 3-10. Wind shifts in direction during a typical day.

Caution: Do not place your faith, or your estimate of the coming weather, on a single indicator. Look at a combination of several factors before you draw any conclusion. Here we have mentioned a combination of changes in wind direction and velocity along with changes in barometer, temperature, and cloud cover. All of these factors when seen together become a strong indication of advancing poor weather conditions as we shall see in Chapters 10 and 11.

Just as backing wind usually precedes poor weather, *veering wind* precedes improvement. A stormy easterly wind may shift through south and southwest, ending at northwest. The temperature will probably drop and the barometer rise. A brisk and gusty wind from the northwest is likely to continue for a day or so in the wake of the storm. The winds will then decrease and become steadier as well.

To help you interpret the meaning of wind directions combined with barometer readings and changes, a Wind Barometer Table developed by the Weather Service is given in Appendix V. This Table indicates the usual consequences of various combinations of barometer readings and true wind directions. *Do not confuse the TRUE and APPARENT wind as explained above.*

HUMIDITY—HYGROMETER OR SLING PSYCHROMETER

Humidity is the amount of water, in the form of vapor, that is mixed with the air. You have most commonly heard it measured and expressed in terms of *relative humidity*. Relative humidity is the amount of water vapor actually contained in a volume of air "relative" to the maximum amount it could possibly hold in vapor form.

Relative humidity is reported as a percentage. Relative humidity when measured as 60 percent means that the air contains 60 percent as much water as air at the present temperature and pressure is capable of holding. Here again, you see, we are involved with the inseparable nature of various factors that together go to make up what we call "weather."

RELATIVE HUMIDITY AND TEMPERATURE

On an absolute basis, warm air can carry in suspension more water vapor than cold air. Let's suppose you've got two cubes of air measuring one cubic foot each. Both are at 75 percent relative humidity meaning both contain 75 percent as much water vapor as they can hold. However, one is at a temperature of 40°F and the other is at 80°F. If all the water vapor in each cube were now to be condensed into liquid water and placed in separate containers, the one from the 80°F cube would contain considerably more liquid than the one from the 40°F cube even though both cubes were at the same 75 percent relative humidity.

Now let's say that the relative humidity of both cubes of air is increased to 100 percent. At this point the air in both cubes is termed "saturated." In absolute terms the actual amount of water in the two cubes is very different, but one critical characteristic is the same: *at this point neither cube of air can absorb any more water vapor.*

If the temperature of either cube goes up, its relative humidity will go down, making it capable of absorbing more water vapor. If the temperature of either cube goes down, it will be unable to retain all the water vapor it already has. Some must condense as liquid water.

In the atmosphere when saturated air (air at 100 percent relative humidity) is further cooled, its water vapor starts to condense as fog, rain, or snow. The temperature at which condensation occurs is termed the *dew point* temperature. This temperature is critical to the prediction of that sailor's nightmare—fog—which is discussed in Chapter 5.

HYGROMETER

The hygrometer measures relative humidity by the amount that humidity lengthens or shortens a human hair (the latest in high-tech?) The changes in hair length are amplified in a manner similar to the system used in an aneroid barometer to make them readable on an instrument scale.

A hygrometer, or a recording hydrograph, of the quality used by the Weather Service is an excellent and useful instrument. However, I've had little luck on boats with inexpensive hygrometers.

SLING PSYCHROMETER

An inexpensive, and for our purposes highly accurate, instrument for measuring relative humidity and determining dew point temperature is the sling psychrometer (Fig. 3-11). The example illustrated folds down into a cylinder 8 inches long by 1 inch in diameter so it certainly will not present a storage problem.

It consists of two identical thermometers one of which has an end wrapped with cotton cloth or wicking. Before use, that cloth must be thoroughly saturated with clean, fresh water. The thermometer with the wet cloth on it is called the *wet bulb* thermometer. The other one is the *dry bulb* thermometer.

These two thermometers are mounted in a holder that permits them to be swung rapidly in the air (Fig. 3-11). This causes some of the water on the wet bulb to evaporate. Evaporation is a change in state from liquid to vapor, a process that requires heat. The heat taken away by evaporation lowers the temperature reading of the wet bulb thermometer.

After about one and a half to two minutes of swinging, the wet bulb thermometer will be as low as it is going to get under existing

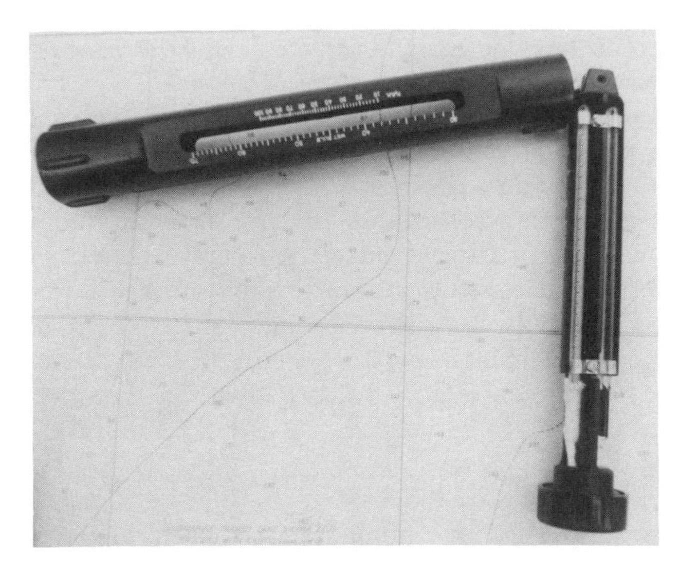

Figure 3-11. Sling psychrometer.

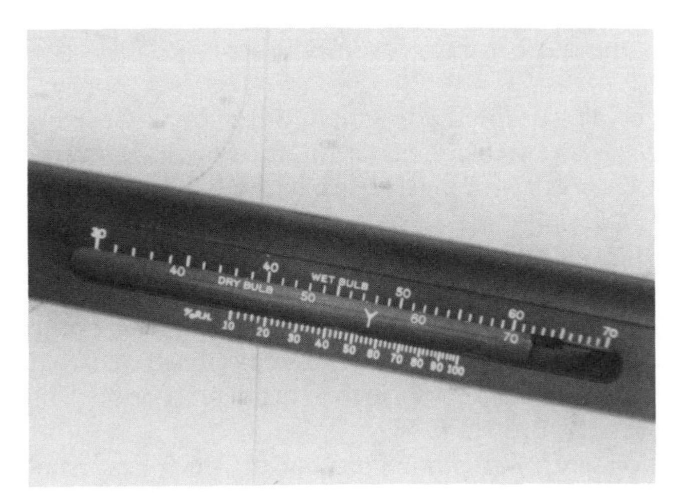

Figure 3-12. Sling psychrometer. Wet bulb reading 60°—dry bulb reading 70° means relative humidity is 55 percent.

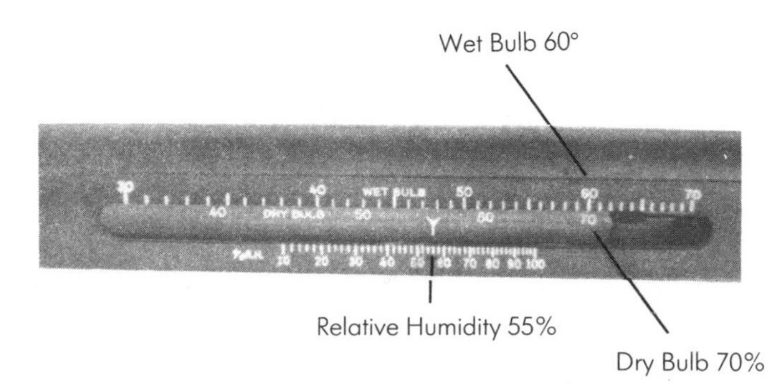

conditions while the dry bulb will be at current air temperature which will be higher. Read the wet bulb first because, as you will quickly see, when you stop swinging the thermometers the wet bulb will start to climb immediately back up toward the same reading as the dry bulb.

On the handle of the instrument, three scales are printed. They are marked "wet bulb," "dry bulb," and "% R.H." (percent Relative Humidity). By moving the scales to match up the wet bulb and dry bulb readings you have taken off the instrument (Fig. 3-12) the arrowhead will mark the present relative humidity. In the illustration, a wet bulb temperature of 60°F with a dry bulb temperature of 70°F results in a relative humidity reading of 55 percent.

In the event you get, or make, a sling psychrometer that does not have its own relative humidity scale, use Appendix Table III. To use this table, enter with the dry bulb temperature in the left-hand column marked "Air temp F." The wet bulb thermometer reading is not used directly. Instead, subtract the wet bulb reading from the dry bulb to get the "Depression of wet bulb thermometer, F," which will be one of the numbers across the top of the table. The two lines thus obtained meet at the relative humidity percentage.

The sling psychrometer is a bit of a nuisance to use, and you need good light to read the thermometers. On the plus side, however, it is an inexpensive instrument, requires only a tiny storage space, and will allow you to keep accurate track of humidity changes.

Relative humidity normally decreases toward the middle of the day as the air warms, and increases during the evening as the air cools. As we shall see in Chapter 5 how much it increases, and how fast, tells us whether fog is likely to form, and if so approximately when.

If you have a sling psychrometer you do not really need a separate marine thermometer such as that shown in Figure 3-3. The dry bulb gives you the current air temperature, and it is probably more accurate. However, a separate bulkhead mounted dial type thermometer is more convenient and easier to read than the mercury column in the psychrometer when all you want to know is the present air temperature.

INSTRUMENT OBSERVATIONS

In order to maximize the value of your own observations as supplements to Weather Service reports and predictions they should be made and recorded systematically and at regular time intervals. Appendix IV is a sample sheet for recording such observations. A Weather Log need not be elaborate to be valuable, but it should include both instrument readings and visual observations as well.

In fair weather, a round of observations every three to six hours is entirely adequate. If you notice clouds thickening, or feel pronounced changes in wind, temperature, or sea condition, it is time to shorten the intervals. When conditions seem to be changing rapidly, cut the observation intervals down to one hour.

If you're not going to bother with a psychrometer, the columns on "Wet Bulb," "Relative Humidity," and "Dew Point" are unnecessary. "Dry Bulb" will be the air temperature as read on any plain thermometer.

Such a log, systematically kept, will be useful in two ways. Today's log gives you guidance as to what today's weather is doing, and consequently clues as to what it is likely to do. Past logs give you an excellent record of previous weather sequences in your immediate sailing area along with what developed from them. Weather sequences tend to repeat. By studying past logs you can develop a broad background on the typical weather changes that occur in the area where you normally sail. Pay particular attention to circumstances that have typically preceded fog, thunderstorms, and other unpleasant weather conditions.

VISUAL OBSERVATIONS

Meaningful visual observation of weather phenomena requires practice, practice, and more practice. However, it is simple, easy, rather fun, and soon becomes second nature. Just as with instrument observations the most important factors with which you are concerned are changes over time.

CLOUDS

Of all the onboard observations you will make, those of clouds will be the most helpful to you in supplementing the information you have received from Weather Service. Clouds provide extremely clear and informative visual signals as to atmospheric conditions and activity. The various cloud types that you will observe while sailing are discussed and illustrated in detail in Chapter 9. The ones you will meet there fall into three fundamental groups: *cumulus* (see Fig. 9-15), *stratus* (see Fig. 9-12), and *cirrus* (see Fig. 9-7).

Cumulus clouds, as explained above, are produced by air from low levels moving upward and cooling. Air that moves up from surface levels is immediately replaced by additional air moving in horizontally at the surface. The sailor feels this horizontal movement as wind.

The small fair-weather cumulus "puff balls" result from weak upward air flow, which has little effect on surface air movement. However, the larger and taller ones are visible indicators of more vigorous vertical air movement. The sailor on the surface will find that when cumulus clouds are growing bigger and higher, they are often accompanied by winds that are becoming stronger and more erratic in both intensity and direction. Beneath cumulus clouds, conditions often change rapidly. Squally winds may alternate with calms while showers alternate with clear and sunny periods.

In contrast, stratus clouds are normally accompanied by winds that remain fairly steady both in intensity and direction. In a sky with stratus clouds overhead changes will occur slowly. If they are thickening and lowering, it will probably rain, but not immediately. When they are thinning and moving higher, clearing may follow, but, again, not right away.

If Weather Service has predicted a major storm, high, thin cirrus clouds are likely to be your first indication that they were right, particularly if they gradually get thicker and cover more and more of the sky. However, in fair weather they may appear for a while, and then dissipate. They may appear and dissipate several times in the course of a day without an appreciable effect on the weather. In any event, when they do indicate changing weather on the way, the change is distant. There'll be no need to batten down hatches for a while yet!

WAVES

Another visible advance indicator of moving weather systems is sea condition—wave height, speed, and direction (see Chapter 6).

Waves are built up by the action of wind on the surface water. In fair weather, a sea that is calm in the early morning may develop waves of 2 to 3 feet by mid to late afternoon when the wind is blowing at 10 to 15 knots. These waves will come from the same direction as the wind that caused them. Typically as the wind dies down at the end of the day the waves die as well.

The longer and harder the wind blows, the larger the waves will be. The strong winds of a storm passing over large areas of water produce giant waves that radiate out over the sea for hundreds of miles from the storm area. As they move they gradually become longer and lower until they finally die out. When you see and feel long low swells from a direction different from the wind, or when there is little or no wind, these swells are showing you the direction of a distant storm. That storm could be one that is predicated to be on its way toward you, it may be one that has already passed, or often those swells are from a storm that will never come anywhere near you.

It is not uncommon to encounter wave trains from two or more directions at once. Quite often you will see and feel waves produced by the local wind along with swells from a second direction. If you've been following the weather reports, you may know where those swells are coming from, and why. If not, it might be worthwhile to find out.

VISIBILITY

The distance of visibility at sea in a small boat is difficult to determine accurately. The clarity of the horizon, or lack of it, is one fairly useful gauge. When visibility is ten miles or over, the horizon is a very sharp, clear line, and sea and sky colors are distinctly separate. When visibility is down to about five miles, the horizon is indistinct and fuzzy. Sea and sky colors are graying.

The causes of changes in visibility are so varied that the distance of visibility at any particular time is not going to be a great help to you in weather forecasting. Poor visibility is produced by both natural causes and by manmade pollution. However, limited visibility under cover of stratus-type clouds is an indicator of stability. Weather changes will be slow. With conditions stable and winds light, it is always well to consider the possibility of fog.

Extremely clear, cool air with scattered small cumulus clouds following a rainy period usually signals the start of a spell of good weather. Very good visibility, but with large towering cumulus clouds moving in, signals a likelihood of sudden squalls, gusty winds, and generally unstable conditions.

In order to draw any meaningful indications for the future from present visibility conditions, one must view them in context with our other indicators: clouds, winds, temperature, and atmospheric pressure.

LOCAL FORECASTING

Weather forecasting for a particular locality on a particular day starts with a basic understanding of the average weather during that season of the year. Start by acquainting yourself with the available data for the area in which you operate. Chapter 4 will provide a beginning. There are several additional sources for this information.

COAST PILOTS

One is the volume of the Coast Pilot that includes your area. These books are published by National Ocean Survey, the same agency that produces your charts. There are nine volumes altogether; however, it is unlikely that you will need more than one. A list of the books and the areas each covers is shown in Figure 3-13. There are four for the U.S. East Coast, one for the U.S. West Coast and Hawaii, one for the Gulf Coast, Puerto Rico, and the Virgin Islands, one for the Great Lakes, and two for Alaska.

You should have the Coast Pilot for your area aboard for navigational purposes anyway. It provides a vast amount of information impossible to include on the charts. In addition, it gives summaries of average weather conditions as well as various types of unusual weather that have been known to occur, or recur in the area from time to time.

PILOT CHARTS

You should also consider getting the Pilot Charts that cover your operating area. That means either a set for the North Atlantic or the North Pacific. A set is twelve, one for each month of the year. These charts are published by the DMA (Defense Mapping Agency) in Washington, and are available from many of the same dealers who handle nautical charts. Each chart provides information on prevailing wind directions and velocities, ocean current directions and velocities, calms, fogs, temperature and barometric pressure averages, and a description of average wind and weather conditions for that month.

TOPOGRAPHY AND LOCAL KNOWLEDGE

Most boating, sailing, and fishing is done in coastal waters where nearby land features often significantly affect local weather. Mountains, deep valleys, forested areas, desert areas lacking in vegetation, or smoke producing industrial developments, all may produce highly localized weather phenomena.

Your local experience plus that of other knowledgeable boatmen in the area will tell you the normal time, direction, and strength of the local sea breeze / land breeze cycle. The wind, temperature, humidity, and cloud conditions that often precede formation of thunderstorms, as well as their usual tracks, are often known locally.

LIMITS OF UNITED STATES COAST PILOTS

ATLANTIC COAST
1. Eastport to Cape Cod
2. Cape Cod to Sandy Hook
3. Sandy Hook to Cape Henry
4. Cape Henry to Key West
5. Gulf of Mexico, Puerto Rico
and Virgin Islands

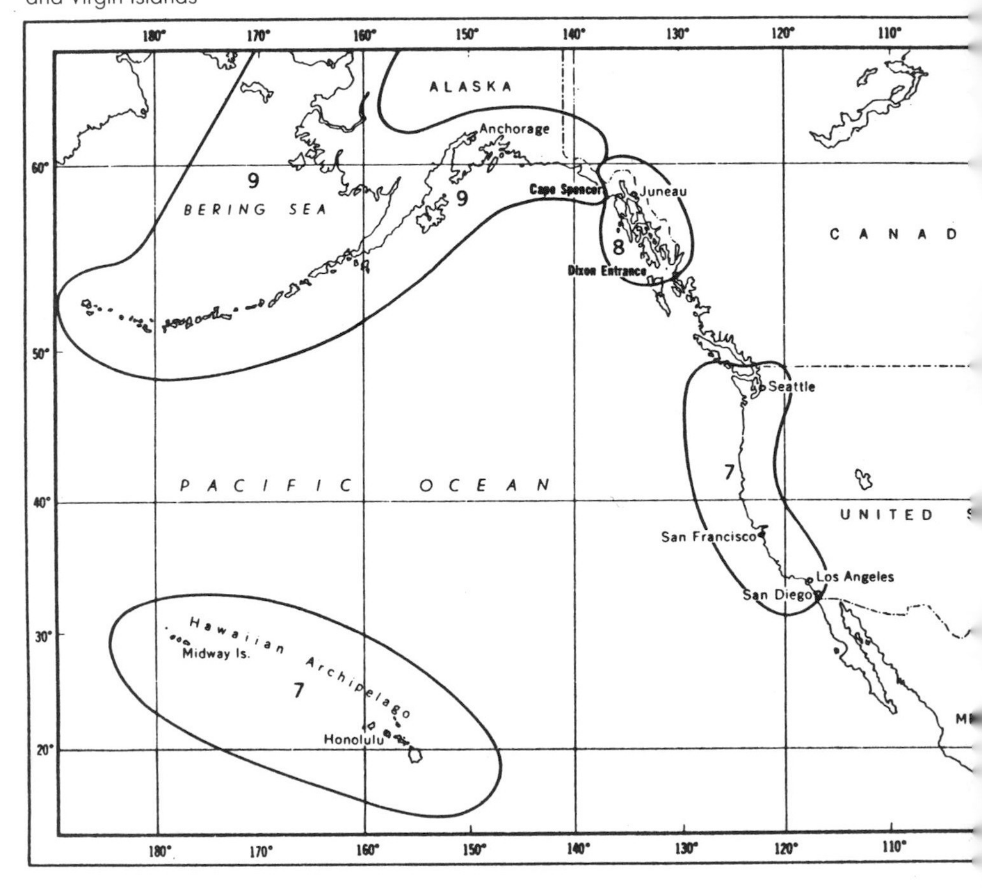

Fog and smog producing conditions are commonly familiar to those in the vicinity as well.

THE LOCAL FORECAST

To make your local weather forecast, you will do, on a small scale, the same thing that is done by the forecasters at Weather Service. First review all of the information available to you up to the minute:

1. The latest available forecast from Weather Service, preferably the current VHF broadcast.

GREAT LAKES
6. The Lakes and their Connecting Waterways

PACIFIC COAST
7. California, Oregon, Washington, and Hawaii
8. Alaska—Dixon Entrance to Cape Spencer
9. Alaska—Cape Spencer to Beaufort Sea

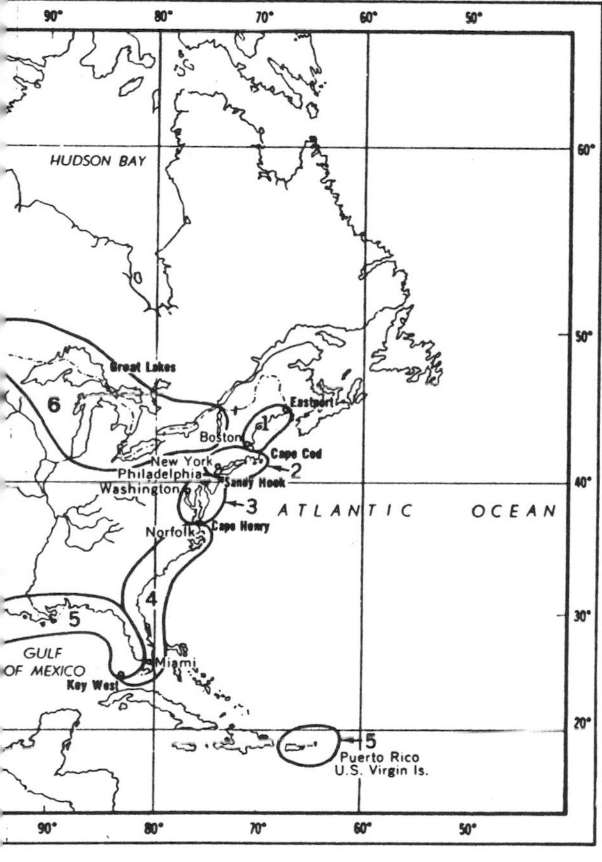

Figure 3-13. Map of coast pilot areas.

2. Your knowledge of the mechanics of weather as discussed in later chapters.

3. Your own record of both instrument and visual observations over the last 6 to 12 hours.

4. The results of your background study of normal and known abnormal conditions in the general area at this season of the year.

By following conditions over several days you will develop a feel for how the weather is presently moving. Your forecast now is going to be whatever is reasonable to expect in light of all available information.

Looking at what you have observed plus the observations, summaries, and forecasts of Weather Service, make a forecast for the next 24 hours and write it down. Next day verify it against what actually happened. Right or wrong, for practice, go ahead and do the same thing again. If you were substantially right—fine! If not, review all the information on which you based your forecast to see if you missed some significant indication.

If your results are not outstanding the first few times, do not be discouraged. In spite of all the scientific equipment and procedures involved, weather forecasting is partly a science but also partly an art. As with other arts, skill follows practice.

4

TYPICAL SEASONAL WEATHER IN MAJOR U.S. BOATING AREAS

Obviously this chapter is *not* going to explain how to predict what the weather will be on Long Island Sound next July 18, or on San Francisco Bay next September 23. The weather to be expected on any particular day at any specific place cannot, as yet, be predicted with any useful degree of accuracy more than a few days in advance.

However, records kept over many years by the Weather Service, and before it by various private individuals, give us a very reliable statistical history of the weather that has occurred at different seasons of the year at a great many locations all over the country. Studying these records we find that by differences in climate and typical seasonal weather, the major boating waters of the continental U.S. break down into seven fairly distinct areas (Fig. 4-1). We also have many U.S. boaters interested in sailing to, or chartering in, the tropics. Therefore, normal seasonal weather is described for the two U.S. tropical areas:

8. Puerto Rico and the Virgin Islands
9. Hawaiian Islands

NORTHEAST ATLANTIC COAST—NORTHERN MAINE TO NEW JERSEY

This area lies in the heart of the belt of the prevailing westerly winds. Its latitude places it directly in the path of frequent large extratropical

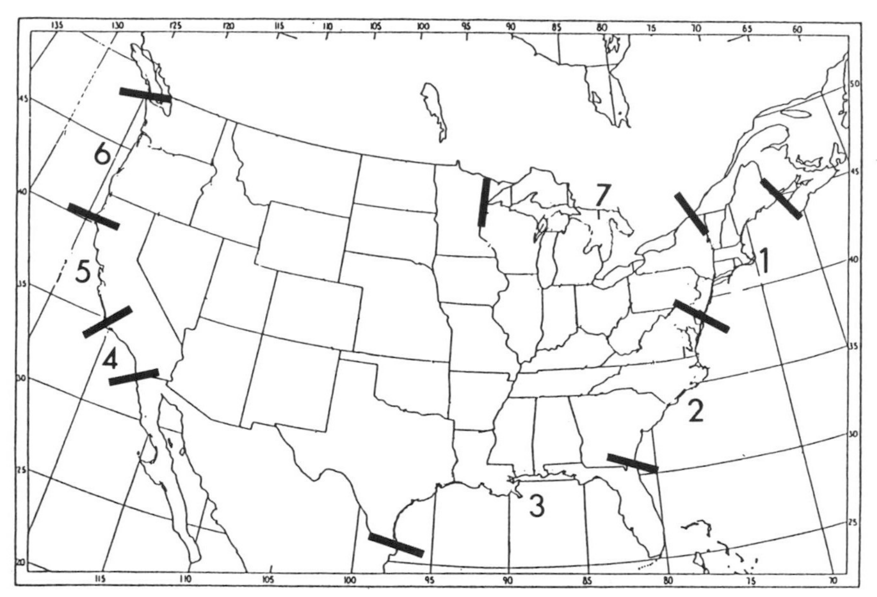

Figure 4-1. Major boating areas of the U.S. 1. Northeast Atlantic Coast—Northern Maine to New Jersey; 2. Middle Atlantic Coast—New Jersey to South Carolina; 3. South Atlantic and Gulf Coasts—Florida, the Keys, and Gulf Coast to southern Texas; 4. South Pacific Coast—Mexican border to Point Conception; 5. Northern California—Point Conception to Cape Mendocino; 6. North Pacific Coast—Cape Mendocino to Canadian border; 7. Great Lakes. Not illustrated: Puerto Rico and U.S. Virgin Islands; Hawaiian Islands.

cyclonic storms (see Chapter 11). It is on the lee side of a huge continental land mass. Its major weather disturbances result from cold continental polar air clashing with warm maritime tropical air (see Chapter 10). At sea, the cold Labrador Current flows southward close to the coast while further offshore the warm Gulf Stream flows northward. The temperature contrast between these currents causes considerable sea fog off this coast.

Cyclonic storms pass regularly from west to east during the cooler months—from mid-fall to mid-spring—embedded in the prevailing westerly air flow. The frontal passages associated with these storms cause abrupt changes in wind force and direction, temperature, and cloud cover. There is a common saying in New England: "If you don't like the weather wait a few minutes—it'll change."

During the warmer months, the Bermuda High, a large stationary area of high pressure over the mid-Atlantic, moves north and intensifies. Coastal winds become mostly west to southwest, and cyclonic storms sharply decrease in number and intensity. These summer winds bring in air that is moist, and warmer than the coastal water. Fog often forms as the air moves out over the cooler sea.

During late summer or early fall, West Indian hurricanes (see Chapter 12) occasionally strike along these coasts. From August to October are the most likely months for them to strike although the full season in the West Indies for these storms is all the way from June to mid-November.

WINDS

From mid-fall to mid-spring the prevailing winds range between west and north. A period of variable winds follows until the summer pattern of west to southwest winds is established from June to September. The strongest winds are from December to March. The weakest blow from May to August. The direction of the summer winds is far more stable than that of the winter winds because of the wide variation in wind directions accompanying winter cyclonic storm systems (see Chapter 11).

Summer winds average in velocity between 10 and 15 knots. Along the coast on hot summer days the land breeze / sea breeze (see Fig. 3-9) sequence often relieves the heat of the afternoon for a distance of up to 10 miles inland. The quiet of hot summer days may be disturbed by occasional severe thunderstorms (see Chapter 12). These generally occur in the afternoon or at night.

On large relatively shallow areas such as Long Island Sound, uncomfortably choppy waves can build up rapidly with a suddenly increasing wind. Being isolated from the swells of open ocean, Long Island Sound and other similar areas are likely to have no wave system at all moving on them during periods of light airs. With no existing swells to interfere, a brisk new wind can quickly develop a very lively, choppy wave pattern.

Winter winds average between 15 and 20 knots except that winds of gale force are frequent during winter cyclonic storms. Winter winds may come in from any direction during the passage of a cyclonic storm system, and at widely varying velocities.

The worst weather conditions encountered in New England are the winter "nor'easters" that accompany advancing cyclonic storms. The nor'easters are strong when a low pressure storm center lies somewhat south of New England (Fig. 4-2). Wind velocities over the sea during these storms are normally greater than over land.

CLOUD COVER

Overcast conditions, meaning 80 percent or more of the sky covered by clouds, can be expected in this area about 50 to 60 percent of the time in winter. In summer, conditions improve considerably, with overcasts occurring only about 30 to 40 percent of the time. Com-

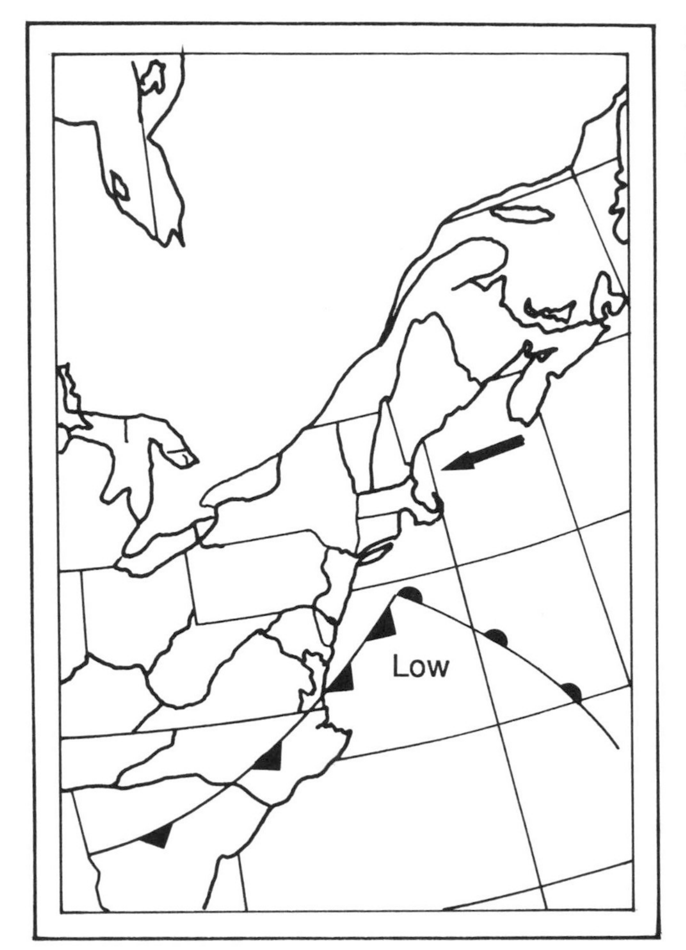

Figure 4-2. The "nor'easter" blows into New England when a winter cyclonic storm center lies to the south.

pletely clear skies are uncommon in this section, but fair skies with about 20 percent cloud cover are seen at least 25 percent of the time.

PRECIPITATION

Precipitation in the form of rain or snow totals approximately 2.5 to 4.5 inches per month. The annual total of somewhat over 40 inches is distributed fairly evenly over the year. Snow may be encountered any time from November to April normally associated with the passage of cyclonic storm systems. Summer rains, in the absence of cyclonic storms, fall from small localized squalls and thunderstorms.

VISIBILITY

Poor visibility at sea in this area is most commonly a consequence of advection fogs which are a summer phenomenon. These summer fogs form at sea over the cold inshore current blanketing the offshore fisherman, while over the warmer coastal bays and harbors it remains clear. At sea summer fogs can be very persistent, sometimes lasting for weeks. In winter the air over the sea is warmer than the land, so when that air blows ashore winter advection fogs form over the land.

MIDDLE ATLANTIC COAST

NEW JERSEY TO SOUTH CAROLINA

The land along the Middle Atlantic Coast is low and flat. Innumerable bays and coves, large and small, indent the coast. A vast number of islands and islets dot the waters close to shore. These form a protected inland waterway along most of this coast. In many instances the bays and sounds are large enough to cause significant local weather variations. The Chesapeake and Delaware Bays are major examples.

Inland, the ground gradually rises, culminating in the Appalachian Mountain Range. This range, while well away from the sea, forms a partial barrier to weather systems moving toward the Middle Atlantic seaboard from the interior. Winter cyclonic storms are modified and deflected by these mountains.

The cold Labrador Current that cools the North Atlantic Coast turns east, but a narrow cool countercurrent still skirts the coast from Cape Hatteras to North Florida. The Gulf Stream is thus much closer to the Middle Atlantic Coast than it is to the North Atlantic Coast. This and the decreasing latitude result in milder winters although some snow and freezing temperatures are to be expected during a normal winter.

In summer, the entire coast comes under the influence of the strong Bermuda High which blocks the passage of continental cyclonic storm systems. For weeks at a time, summer weather in this area does not change significantly. Local instability, showers, and thunderstorms are frequent as the coast lies along the west side of a large stationary high pressure system (see Fig. 10-5). High temperatures and humidities, along with low wind velocities mark the summer season here.

The Appalachian Mountains and the Bermuda High together effectively keep most cyclonic storm systems out of this area during the summer. However, the entire coast is exposed to the passage of West Indian hurricanes (see Chapter 12). These move west and north with the trade winds through the Caribbean until they reach about latitude 25°N where they start to curve north and east. Depending on where this happens, they may strike anywhere along the U.S. Gulf or Atlantic coasts. While the usual season runs from June to November, most of the hits along these coasts have been during the period from August to October.

WINDS

Winds during the cooler months, from October to March, are predominantly from the north through west quadrant with some southwest winds in the more southerly portion. Winter storms bring easterly winds, but seldom with the intensity of the New England "northeaster" except along the Virginia Capes particularly from Cape Henry to Cape Hatteras.

In spring, coastal winds move more and more to the southwest as the Bermuda High starts to move north and west. In the southern portion, the weather becomes warm and humid as the cyclonic storms decrease. Some frontal activity continues in the northern portion bringing spring rains alternating with fine, mild weather.

In summer, the influence of the Bermuda High dominates the entire coast. Weather is warm and humid with low wind velocities primarily from the southwest. In many areas the daily land breeze / sea breeze cycle takes over. Local summer showers and thunderstorms punctuate a stable weather pattern that changes little for weeks at a time. Late summer and early fall may also be abruptly disrupted by the arrival of tropical hurricanes anywhere along this coast.

In the fall, the Bermuda High gradually retreats south and east allowing cyclonic frontal systems again to enter the area. The transition to the winter wind and weather pattern starts in the north and gradually works southward.

CLOUD COVER

Completely clear skies are extremely rare. Cloudiness usually ranges somewhere between 35 and 70 percent of the sky. Cumulus clouds predominate throughout the area. In summer, this results from the instability of the air on the west side of the stationary Bermuda High (see Fig. 10-5). In winter, cumulus clouds come from unstable air associated with passing cyclonic systems.

Maximum cloudiness occurs during the winter in the northern part of the area due to the passage of many cyclonic storm systems. These rarely affect the southern sections because they are blocked by the Appalachian Mountains.

PRECIPITATION

Precipitation is heavier in the south than in the north. In the north, it averages 35 to 45 inches per year. Toward the south, it ranges as high as 45 to 60 inches per year.

Heaviest precipitation in the north is 4 to 6 inches per month during July and August with the least being 2 to 3 inches in October and November. In the south, the heaviest months are September and October with the lightest being February. The arrival of a tropical hurricane totally disrupts the normal pattern since as much as 9 to 15 inches of rain may fall in a single 24-hour period.

In the northern section, snow may be encountered any time from December to March on a few days a month. However, toward the south, snow is not a significant hazard. The area is essentially snow free year round.

VISIBILITY

Visibility is generally good here with fog being the principal restriction encountered. The worst fogs occur in winter in conjunction with frequent air mass changes and cyclonic activity. Restricted visibility is usually worse at night and in the early morning. Fogs decrease over the area as you move from north to south.

Along the coast, radiation fogs frequently form after sunset and may severely restrict harbor activities, but they generally do not extend any significant distance seaward. Summer sea fogs sometimes drift ashore on hot days, and may persist for several hours, seriously hampering maritime activities.

SOUTH ATLANTIC AND GULF COASTS

FLORIDA, THE KEYS, AND THE GULF COAST TO SOUTHERN TEXAS

Coastal terrain varies from flat to gently rolling. It is indented by many bays large and small. Numerous offshore islands form a protected inland waterway along most of the coast from the northern Atlantic side of Florida all the way around the Gulf to the Texas-Mexico border.

The general climate in this area ranges from humid subtropical in southern Florida and southern Texas to a more variable, but still warm marine climate in the more northern parts of Florida and the long sweeping curve of the Gulf Coast.

During the warmer months of the year, warm moist air from the Gulf of Mexico flows north, moderating the climate of the Gulf Coast. Along the Atlantic side of Florida, the warm Gulf Stream stabilizes the climate through most of the year. Winters are mild. Summers, while warm and humid, are not extreme.

During late summer and fall, the entire area lies in the path of West Indian hurricanes that have turned north from the Caribbean. August and September are the months when hurricanes are most likely to hit along these coasts.

WINDS

Over Florida and the Gulf Coast south of latitude 30° N, easterly winds are prominent throughout the year. Along the central part of the Gulf Coast, the winds vary with the passage of the large continental cyclonic weather systems. Between 30 to 40 percent of midwinter coastal winds are from the northern quadrant while 40 to 50 percent of summer winds are southerly. Light to moderate winds are normal in this area, and the daily land breeze / sea breeze cycle commonly prevails.

In winter, a number of polar air masses (see Chapter 10) penetrate through to the Gulf. During a typical year some 15 to 20 of these disturbances, locally termed "northers," bring strong northerly winds and lowered temperatures to this area.

Strong winds and unpleasant conditions are generally associated with storms having centers of low atmospheric pressure. Extratropical cyclones, hurricanes, thunderstorms, and tornados all share this characteristic (see Chapters 11 and 12). The Gulf area "norther" differs (Fig. 4-3) in that it blows out of a high pressure anticyclone.

A mild norther will bring winds of at least 20 knots. A severe one may blow as high as 50 knots. In any one year up to 6 northers might be severe, and these will probably occur in December or January. A mild norther will usually last about a day and a half, but a severe one may last 3 or 4 days.

CLOUD COVER

On average, clouds cover between 35 and 60 percent of the sky throughout this area. Completely clear skies are very rare. Gray, overcast days occasionally occur in winter, but very seldom in summer. The clearest month is usually October while the cloudiest skies are during the period between December and March.

Much of the summer cloud cover consists of fair-weather cumulus clouds (see Figs. 9-15a and b). Winter northers may bring any and all

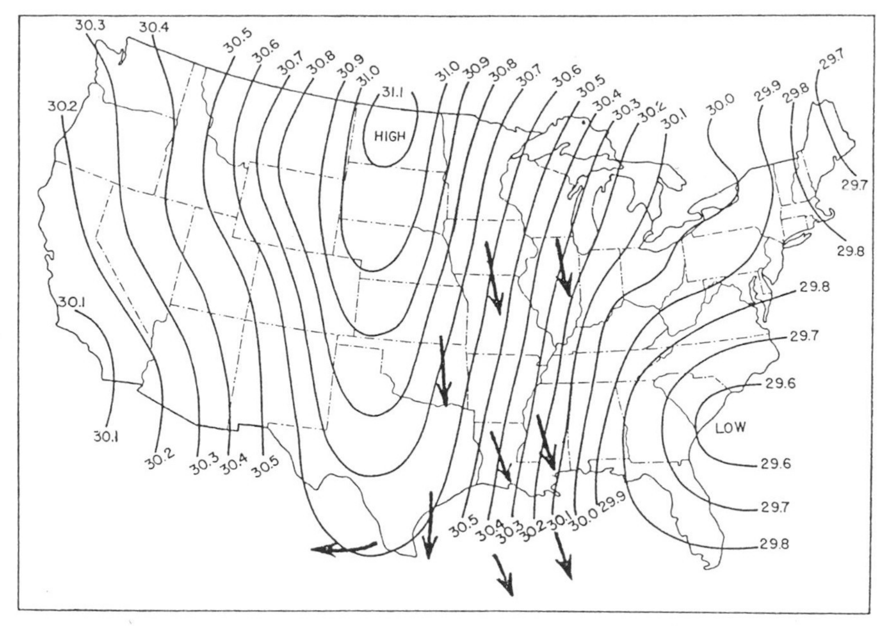

Figure 4-3. Conditions resulting in a strong "norther" along the Gulf Coast. Courtesy: U.S. Weather Service.

of the cloud types associated with cyclonic frontal passages (see Chapter 10).

PRECIPITATION

Along the Gulf Coast precipitation accompanies the periodic northers and cyclonic systems of late fall, winter, and spring. Thunderstorms and showers are frequent during summer and early fall when cyclonic storms are rare. By far the heaviest rains, however, are associated with the tropical storms or hurricanes that occur in August, September, and October.

Monthly totals range from 2 to 6 inches except in the event of a hurricane. As has been mentioned a hurricane can drop as much as 9 to 15 inches of rain here in a single day.

VISIBILITY

Throughout this entire area visibility is generally good. At sea, fog is virtually nonexistent. Along the Gulf Coast from November to April, warm, moist air blowing slowly in over chilled land surfaces occasionally forms low fogs over the land. They sometimes extend over bays and harbors. These fogs form with southerly winds and dissipate with northerlies. Over Florida, coastal fogs are very infrequent except in the vicinity of Tampa during the winter.

SOUTH PACIFIC COAST

MEXICAN BORDER TO POINT CONCEPTION

The coast trends east from Point Conception to Santa Barbara then curves southeast to the Mexican Border just south of San Diego. Numerous sandy beaches alternate with low cliffs. It is generally exposed to the prevailing westerly winds and swells, with few harbors and protected anchorages. Inland, a range of mountains approximately parallels the line of the coast. A group of islands known collectively as the Channel Islands lies at varying distances off this coast. Of these Santa Catalina south of Los Angeles is by far the best known and the most popular boating destination.

This coastal climate is mild subtropical. Major weather disturbances are infrequent as are temperature extremes. Winds are generally moderate. Rainfall is sparse.

Major factors influencing the weather in this area are the movements of the semipermanent Pacific high pressure system, the cold California Current, and the thermal low pressure area that forms over the inland deserts. The Pacific High dominates the weather along the entire West Coast much as the Bermuda High dominates the East Coast.

The normal presence of the Pacific High off the coast coupled with the normal inland thermal low pressure produces an atmospheric pressure difference that results in a flow of air from the ocean in over the coast. The coastal water of the California Current is cold. This cold water cools a low layer of the air flowing in from the ocean. This low layer of cool air, called the *marine layer,* underlying warmer air above results in a persistent temperature inversion (see Chapter 8).

Normally atmospheric temperature decreases steadily with altitude. When an inversion exists, air temperature increases with altitude for some distance rather than decreasing. In the case of a low inversion (Fig. 4-4a) the area of increasing temperature starts at ground level. In the case of an elevated inversion, air temperature decreases for some distance (Fig. 4-4b), then increases, and then finally again decreases. The cold air in the low layer of the inversion, being heavier than the warm air above it, forms a lid, stopping vertical air move-

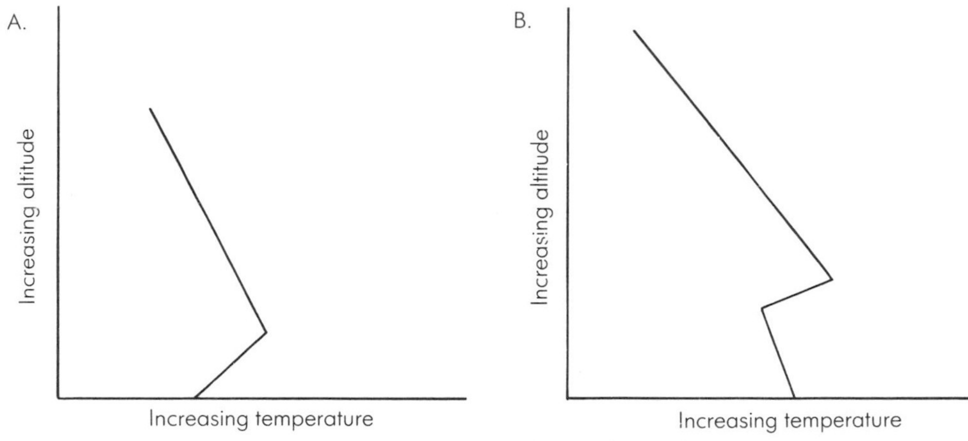

Figure 4-4. Temperature inversions. a. Low inversion often resulting in fog or smog. b. Elevated inversion produces low stratus clouds.

ments. A low inversion often produces fog or haze along these coasts while an elevated inversion commonly results in low stratus clouds. The infamous Los Angeles smog is one unpleasant result of this inversion layer.

Between June and September, storms in these waters are almost unknown. Frontal weather begins to penetrate here during October. The number and intensity of frontal storms gradually increases through December. From January through March four to six frontal storms per month of from one to two days duration are typical. During April and May frontal disturbances become milder and less frequent, disappearing by June.

Tropical cyclones, known in Mexico as *chubascos,* form in summer and fall off the west coast of Mexico. Of the ones that reach hurricane intensity, very few get this far north. When they do, they have generally lost full intensity, although from time to time they bring abnormally high surf particularly on south facing beaches. They may also bring rain or thunderstorms.

WINDS

Light to moderate winds predominate in this area varying from mostly northwest in summer to west and southwest in winter. A common fair weather pattern is light variable winds night and morning hours becoming westerly 10 to 16 knots during the afternoon. Gales are infrequent, occurring between mid-fall and late spring usually in conjunction with the passage of a frontal system. From November on through the winter, fairly vigorous frontal systems pass through, usually bringing winds of 25 to 35 knots.

Two types of wind conditions occur here that are departures from the normal day to day wind patterns, and are also peculiar to Southern California. These are the Santa Ana winds, and the Catalina Eddy.

SANTA ANA WINDS

These are dry northeast winds that blow down out of the mountains onto the coast and coastal waters anywhere from Point Conception to the Mexican Border and below. They most frequently occur during the period from November through January, but may occur at any time.

A Santa Ana (Fig. 4-5) blows when surface atmospheric pressures become higher over Nevada and Utah than over Southern California. Conditions become favorable for this to happen when a low pressure storm has passed inland over Northern California followed by a mass of high pressure (see Chapters 10 and 11). Air then flows from the high pressure area inland down through the mountains to the lower pressure area along the coast.

As it comes down from the mountains, the descending air warms at the rate of 5.5 degrees per 1000 feet. When it arrives at the coastal plain, if it is warmer than the air there (a warm Santa Ana) it rides up over the cooler air, localizing its effects to coastal canyons. When, after descending, it is still cooler than the coastal air, it pushes the coastal air out to sea. This cold Santa Ana often brings very sudden strong winds as high as 30 to 40 knots or more to large sections of the coast and coastal waters, particularly toward the east end of the Santa Barbara Channel. Since these winds often come up without warning at night, they can be very dangerous to small boats lying at anchor either in mainland harbors or in the vicinity of the Channel Islands. In this area during clear, dry weather with a high or rapidly rising barometer, be alert for a possible Santa Ana.

THE CATALINA EDDY

As mentioned earlier, the Pacific High dominates the weather patterns along the U.S. West Coast. In summer, this high pressure area increases in intensity and moves north. In consequence an increase occurs in the atmospheric pressure difference between the Pacific High and the summer thermal low pressure area that develops over the inland southwestern deserts. A persistent northwest summer wind (see Fig. 10-5) along the California coast results.

Figure 4-5. Typical pressure pattern that produces Santa Ana winds.

Just above Point Conception, the coast makes a sharp bend to the east. When a fast wind (25 to 35 knots) is blowing down the coast, it tends to turn east following the coastline (Fig. 4-6). This easterly swirl is the "Catalina Eddy." It may simply turn in toward the coast or, particularly at night, it may form a completely circular pattern over the coastal waters. The night land breeze, part of the daily land breeze / sea breeze cycle, reinforces the circular eddy pattern. The name "Catalina Eddy" derives from the fact that the center of the completed circular eddy pattern is often located over the offshore island of Catalina.

Due to the influence of the Pacific High, the normal summer winds in Southern California (see Fig. 10-5) are northwest. However, during a Catalina Eddy they turn southwest in the area south of the eddy from about San Pedro and Long Beach on down.

The small boat skipper has no way of telling that an eddy is forming based on what he can see from his boat. However, a weather broadcast giving a forecast similar to the following would give him a clue: "Northwest winds 20 to 30 knots over outer channel waters today and tomorrow. Near the coast southeast winds 5 to 10 knots during night and morning hours becoming southwest 10 to 18 knots in the afternoon. Coastal low clouds nights and mornings becoming mostly sunny in afternoons."

The development of an eddy brings with it an increase in fog or low clouds. If no fog exists when the eddy starts, fog formation is likely. If it is already foggy, that fog is likely to increase, then lift to form a layer of low stratus clouds. If low stratus clouds already exist, they are likely to move higher improving visibility below.

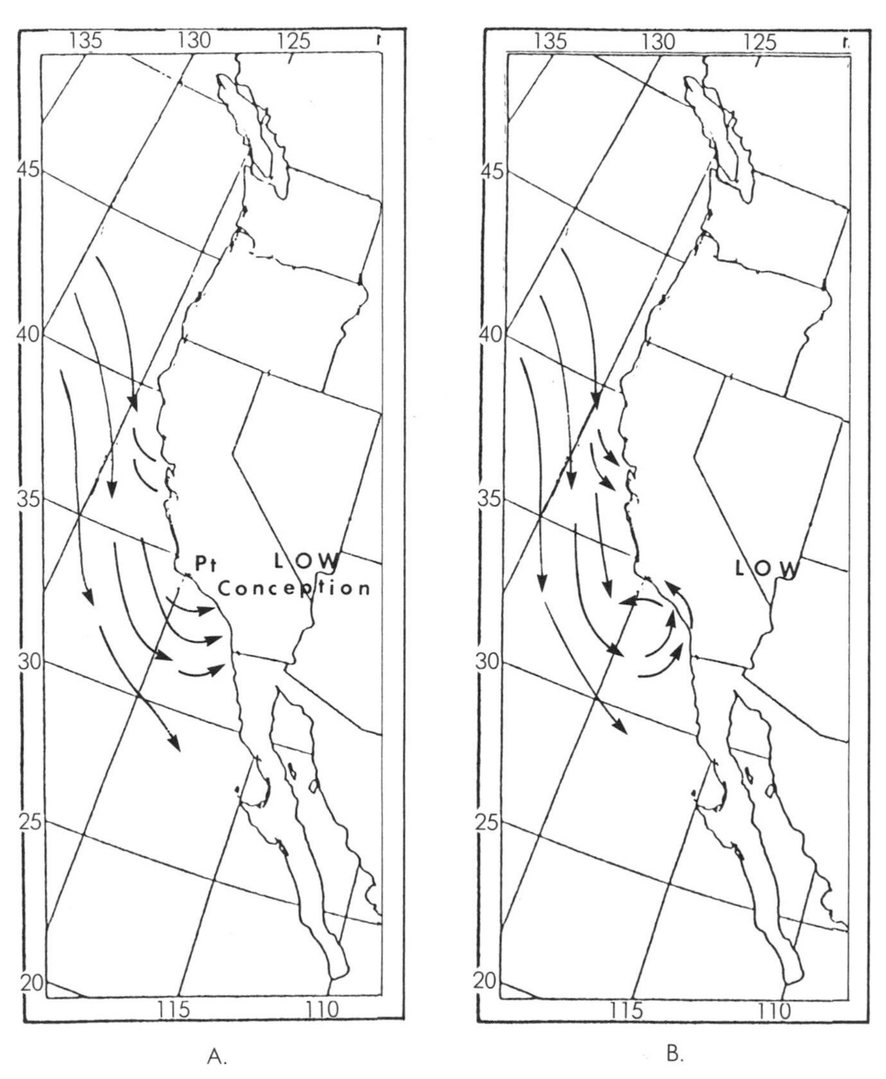

Figure 4-6. Wind circulation called "Catalina Eddy." A. Wind stream from the north-northwest moving down the California coast turns east toward the land after rounding Point Conception. B. A nearly closed "eddy" is formed during the night when the effect of the night land breeze obstructs the flow of air toward the coast and deflects it to the north and northwest.

CLOUD COVER

In consequence of the persistent temperature inversion along this coast, the typical cloud forms are stratus types. Late night and early morning stratus clearing by mid-morning is a common sequence here. The passage of frontal weather brings, of course, the succession of clouds normally associated with the type of front involved (see Chapter 10).

Late May, June, and early July bring the most persistent low stratus overcasts often lasting all day. Late fall through April brings the clearest skies except during the passage of frontal disturbances.

PRECIPITATION

Rainfall along this coast is sparse, coming mostly in conjunction with the frontal weather that comes between November and April. Most of the normal annual total of between 12 and 14 inches falls between December and February. In a typical year, considerable snow falls on the mountains just a few miles inland, however, none falls on the coast or the coastal waters.

VISIBILITY

Due to the persistent temperature inversion in this area, both fog and haze often restrict visibility along this coast. They are aided by the infamous Southern California smog. When the inversion is low, the result is often fog or low stratus clouds at night, gradually clearing but with haze persisting during the day. With evening cooling, the stratus clouds or fog reform.

Fog formation usually occurs here at night. However, occasionally the day sea breeze brings an early afternoon advection fog bank in over the coast. These fogs comes in fast enough to catch an unwary boatman unprepared. Fortunately, when they come in fast they usually do not last long, and dissipate quite rapidly. If caught in one of these fast moving fogs the best thing to do is to heave to or anchor until it lifts.

NORTHERN CALIFORNIA

POINT CONCEPTION TO CAPE MENDOCINO

Except for very occasional bays and harbors this coast is mostly steep and rugged with numerous rocks and shoals close inshore. Other than at San Francisco and Monterey Bay, harbors providing shelter from the prevailing westerly winds and seas are virtually nonexistent.

Weather along this stretch of coast is cool, damp, and foggy in summer. Winter is damp as well but temperatures remain mild. In summer, the strong Pacific High forces most storms and fronts to the north and clear of this area. In winter the high weakens and moves south, allowing storms or fronts to pass through every week or two. Sometimes a series, or family, of storms (see Chapter 11) may bring a prolonged period of strong winds and rains.

The combination of San Francisco Bay, San Pablo Bay, and Suisun Bay, plus smaller connecting bays along with the Sacramento and San Joaquin rivers, constitutes by far the largest seaport area on the U.S. West Coast. It also offers a vast protected area for recreational boating, fishing, and sailing.

While its waters are protected from the heavy seas of the northern Pacific Ocean, the San Francisco Bay shares the general windiness and fogginess of the rest of Northern California.

WINDS

Winds are much stronger in Northern California than in the south. In summer, the Pacific High brings northwest winds that are reinforced by the inland thermal low to the southeast. The result is a sea breeze of up to 20 knots in the afternoon dying down by evening but persisting as late as midnight.

Winter winds from November through February are variable due to the frequent passage of frontal weather systems. Shifts in wind direction and speed are frequent. Extreme winds of 50 knots with gusts to 75 have been recorded. These have usually been out of the southerly quadrant in advance of a cold front.

The windiest time in terms of average wind speed is often spring. The extremes of winter are unlikely but steady winds in the 17 to 28 knot range are likely. Wind direction becomes less variable as well. It steadies more and more on the northwest as the Pacific High moves to its summer position.

CLOUD COVER

The steady, summer northwest winds blowing over the cold California Current brings low stratus clouds and fog to the coast. This condition is prevalent during the early mornings, clearing as the day progresses. The worst months are August and September.

The frontal weather systems of winter bring with them their typical cloud sequences (see Chapter 10). The various stratus types come in with warm fronts, cumulus types with cold fronts. Skies are clearest in winter and spring during the breaks between frontal storms.

PRECIPITATION

Rainfall increases gradually as one moves north along the coast, but is still quite sparse, averaging 17 to 19 inches per year. Most of this falls during the winter frontal storms between November and April.

There is little or no rain during the summer months. Considerable snow falls inland on the mountains, but not on the coast.

VISIBILITY

Sea fog and haze make visibility generally poor during the summer months particularly during August and September. The NW wind from the Pacific High is reinforced as it is drawn in toward the inland thermal low pressure. As it passes over the cold coastal waters, advection fog results. Sometimes these fogs will persist all day. At other times they clear during the middle of the day, and sometimes they lift with the day sea breeze to become low stratus clouds.

Sea fog becomes infrequent during fall and winter, normally occurring only with the warm, moist air flow ahead of a frontal disturbance. In the San Francisco Bay area the most common fall visibility problem is smog. This is made worse when winds are light and a high pressure system has settled in.

Another cause of poor visibility in the Bay Area is called *tule fog*. It gets its name from the fact that it is a radiation fog that forms over marshy areas where tall bullrush type plants called tules grow. Tule fog forms in calm weather particularly during the winter months of December and January. It may be only a few feet deep, but can work up to several hundred feet deep and may persist for several days if the calm holds.

In spring, particularly in March and April, visibility is usually excellent. Winds are stronger and weather disturbances less frequent, reducing the chances of both advection sea fog and radiation land fog.

NORTH PACIFIC COAST

CAPE MENDOCINO TO CANADIAN BORDER

Along this stretch, high mountains closely border the coast. Much of the coast is rugged with many detached rocks menacing navigation. While there are few harbors for deep draft ships, there are many bays, coves, and harbors suitable for small boat use.

The weather is dominated by the movements of the Pacific High and the Aleutian Low. As the prevailing westerly winds reach the coastal mountains moist air is lifted and cooled. Much of the moisture condenses into clouds and rain which falls heavily on the coastal side of the mountains, and decreases on the inland side.

The *maritime* climate along the coast keeps winters mild and summers cool. Moving inland, up the Columbia River for example, the climate becomes more *continental*—warmer in summer and cooler in winter.

WINDS

While temperatures are mild along this coast it is windy. Winter storms bring E through SE to S winds up to gale force, along with frequent rain or occasional snow on up to 15 or 20 days per month. Between storms NW winds are common.

By spring, winds diminish and move to the NW to SW quadrant. In summer, storms have become infrequent and winds have become quite light predominantly from a westerly direction. The heating of the ground inland reinforces the sea breeze during the day. The opposing land breeze produces calms on many nights. During the fall, the storms gradually increase in force and frequency as the weather returns to the winter regime.

CLOUD COVER

With winter storms bringing rain anywhere from 15 to 25 days per month, winter skies are overcast a great deal of the time. Spring and summer brings much fair-to-partly-cloudy weather, but very few completely clear days.

PRECIPITATION

Rainfall gradually increases from south to north along the coast. At Eureka, California, just north of Cape Mendocino, the annual rainfall is about 38 inches. At Astoria, near the mouth of the Columbia River in Oregon, it is up to about 80 inches, and in northern Washington it reaches as much as 104 inches.

At major inland ports, such as Portland and Vancouver on the Columbia River, or Seattle and Tacoma on Puget Sound, annual rainfall drops again to about 38 inches. The coastal mountains have wrung much of the moisture out of the weather systems moving in from the west.

VISIBILITY

Just as everywhere else on the U.S. West Coast, fog is a constant problem. Summer brings frequent advection fogs over the cold coastal waters. With light winds, these fogs may persist for days.

In winter, frontal fogs may accompany the passage of storm systems. Winter is also the time for night radiation fogs that form over land, later drifting out over bays and coastal waters. These usually quickly burn off with the morning sun.

THE GREAT LAKES

The Great Lakes lie in an area over which polar air from the north, and temperate air from the south fight back and forth, producing an immensely wide range of weather conditions. As contrasting air masses meet over the Lakes, the formation of intense, fast-moving cyclonic storm systems (see Chapter 11) is often triggered. However, most of the worst winter storms on the Lakes come in already formed from the W and SW. These are spawned in the Pacific southwest (Arizona / New Mexico), and the mountain and plains states further west.

Another source region for storms on the Great Lakes lies in western Canada. These storms, called "Alberta Lows" are most frequent in October. They are not usually as severe as the ones that originate in the W and SW U.S.

When an eastward moving cyclonic storm center is passing north of you on the lakes, the signs will be a steadily falling barometer, SE to S winds, and constantly lowering stratus clouds followed by drizzle, rain, or snow. As the warm front passes, the rain stops, winds move to the SW, skies clear, and temperature rises. Arrival of the cold front brings banks of cumulus clouds and the wind becomes squally and shifts to W then NW. There may be rain squalls, showers, or frontal thunderstorms (see Chapters 11 and 12).

If you are north of the storm center, the weather changes are slower and less pronounced. Winds gradually back from E through N to NW. The stratus clouds and steady rain, similar to conditions ahead of a warm front, gradually change to cumulus type clouds and squalls, similar to cold front conditions.

In summer, cyclonic storms become less frequent and less severe. However, from May through September, with a frequency of between 5 and 10 days a month, the Lake sailor may encounter a thunderstorm (see Chapter 12). These range from comparatively mild disturbances to violent squalls with torrential rains and wind gusts as high as 100 knots! In addition, thunderstorms may harbor tornados or waterspouts. Along the shores they are most likely to occur during afternoon or evening hours. Over open water they are most likely at night.

From about mid-December to early April, most commercial shipping is suspended because of the combination of ice and poor weather.

When large ships fear to go out, only the foolish will venture out in small boats!

LAKE ONTARIO

Fall brings the strongest navigation season winds with gales out of the NW to SW quadrant likely from October through December. With W and SW winds blowing most of the year, a funneling effect may occur toward the east end of the lake as it narrows. Here, what was a moderate blow in mid-lake can become a dangerous gale. Winds from May through August are usually moderate. Eighty percent of the time they are likely to be sixteen knots or less.

Prolonged periods of rain and fog develop when weather fronts moving toward New York become stalled. Spring brings some advection fogs, usually at their worst in the mornings. Calm, clear fall nights may bring radiation fogs that drift out over the water, only to burn off the following morning.

Rough seas can make up at any time, but are most likely from October to February. Five- to ten-foot waves are common with extreme heights up to nineteen feet. From May to July sea conditions are best. Then for at least half the time waves are less than one foot.

Most of Lake Ontario remains ice-free except during severe winters. Even then significant ice is confined to the east end of the lake, and is mostly gone by the end of March.

LAKE ERIE

Like Ontario, Lake Erie has its strongest sailing season winds in autumn, and its lightest from May through September. It is also oriented so that its greatest length is exposed to the prevailing W to SW winds. Since it is a shallow lake, these winds can quickly raise dangerous, steep, short-period waves. Close to a third of the time, waves are five feet or more. Maximum heights reach fifteen to twenty feet.

Poor visibility is a navigational problem mostly during spring and fall. Spring brings advection fogs over the open waters which drift in over the coasts. Fall radiation fogs drift out from the shores. Fog is more frequent on the north shore than on the south.

The west end of the lake is particularly shallow and freezes rapidly in late fall or early winter. Because of warming temperatures and prevailing winds, the west end starts to clear of ice by mid-March. The east end freezes later, in January, but because the prevailing winds force lake ice toward the east, it can persist at that end through late May after a severe winter.

LAKE HURON

Lake Huron is large, and shaped so that strong winds from any direction will raise rough seas somewhere on the lake. W through NW winds of fall and winter are often the strongest. Waves of ten feet or more will be raised in open waters by a twenty-knot wind, and waves as high as twenty-two feet have been recorded at times.

Winds from a northerly quadrant raise dangerous seas toward the south end of the lake particularly near the southern outlet. Strong easterly or northeasterly blows across the central waters can build high seas along the Michigan shore directly across the main north-south traffic route. Southerly winds stir up dangerous seas in the northern section of the lake. Winds are mostly moderate from late spring through summer except for occasional extremely violent line squalls and thunderstorms.

In spring and early summer, dense fogs caused by warm air moving out over lake water that is still cold form over the open waters. These fogs most often form with winds out of a southerly quadrant, and are thickest during the mornings.

While the central part of the lake stays generally ice free in winter, the shores, bays, ports, and channels in and out of the lake freeze over from December to March or April. The Straits of Mackinac get particularly heavy icing. Ridges as thick as thirty feet have been seen.

LAKE MICHIGAN

The greatest length of Lake Michigan is north-south. Consequently the worst sea conditions are raised at the south end by northerly winds, and in the north end by southerly winds. During the entire navigational season from April to December, winds with a southerly component are prevalent. Northerlies are less common, but frequent in spring. Sailing conditions are worst in October and November when at least a third of the time seas are running between five and ten feet.

Strong winds are infrequent during the summer. Those that do occur are associated with thunderstorms, and are of brief duration. The land breeze / sea breeze sequence becomes common in summer. This means the day breeze is easterly on the west shore, and westerly on the east shore, both reversing at night.

Fog on the shores of Lake Michigan is variable. In spring and summer, advection fogs are more common along the NW shore than elsewhere due to an upwelling of cold water. The Chicago area is subject to smog and haze, but less actual fog than occurs to the NW.

Sunrise or early morning fogs may occur at any season. The mid and later part of the day in spring and summer is generally clear.

Ice coverage on the lake varies from ten percent during a mild winter up to eighty percent in a severe one, but by mid-December the Straits of Mackinac are usually closed, blocking traffic in and out of the lake. By mid-April they are usually open again. While large parts of the central waters of the lake may be clear of ice in mid-winter, the coastal bays and harbors are not.

LAKE SUPERIOR

Summer winds are out of the S to W quadrant. They often shift to NW at the east end of the lake. Speeds are in the ten- to twenty-knot range. Better than half the time wave heights are two feet or less, building to between five feet and ten feet part of the time.

By September, wind speeds and wave heights are increasing. Waves reach five feet or more close to a third of the time. As the fall progresses, winds are more out of the W and N and increasing. Also increasing is the frequency of cyclonic storms. Winter winds continue to increase in force with northerlies and cyclonic storms becoming more common.

Spring winds are variable with gales becoming less frequent. However, at many places April brings the highest mean wind speed of the year. In April, waves of five feet and over are found about a third of the time, dropping to ten and fifteen percent in May followed by another major drop again in June.

Visibility can be poor at any season. In fall, radiation fog, in winter, ice fog, or in spring or fall, advection fog may develop. Along the shores dense fog is mainly a morning phenomenon. During summer, advection fogs may drift onshore; during fall, radiation fog may drift out over the water. In the vicinity of industrial cities, smoke and smog add their few cents worth to the visibility problems.

Winter ice coverage of the central waters of the lake may reach up to eighty or ninety percent. But even when it is far less than that, virtually all bays, harbors, and entrance channels are solidly iced over. From December to mid-April, do not expect ice-free water even when you do get a day of beautiful weather.

PUERTO RICO AND THE VIRGIN ISLANDS

These tropical islands lie well within the belt of the easterly trade winds. Puerto Rico is fairly large, approximately 125 miles long and

35 miles wide. Its hilly interior is mostly surrounded by flat coastal plains. The Virgin Islands are small, the largest, St. Croix, being a little over 20 miles long. While St. Thomas is rather hilly the other Virgin Islands are quite flat.

The climate is warm and humid. The surrounding seas and steady trade winds have a tempering effect, keeping daily temperature ranges small on the Virgin Islands and the coastal areas of Puerto Rico. Daily highs are generally between 82 and 88 degrees year round. Summer highs occasionally get into the 90s. Daily lows vary from the 70s down to 65.

The north coast of Puerto Rico, along with the north coast of St. Thomas and St. Croix, is exposed to the swells of the North Atlantic. Winter "northers" occasionally bring heavy seas for short periods to these areas. Otherwise, seas are generally moderate.

WINDS

The easterly trade winds are extremely dependable both as to direction and force. In winter they blow from the NE shifting to E and SE in summer. They are light to calm during night and early morning, blowing fresh and steady from late morning through the afternoon. The "reinforced trades" of winter bring afternoon winds as high as 18 to 20 knots at times. By late summer, the afternoon maximum drops to the 10- to 15-knot range.

While hurricanes seldom strike these islands, when they do they bring disastrous winds. In one hurricane, the Weather Service at San Juan lost its wind speed indicator after it had recorded 160-mph winds. The occasional passage of hurricanes nearby brings strong winds and high seas.

CLOUD COVER

Fair weather with a constant scattering of small cumulus clouds is typical. Periodically the normal flow of the trade winds is interrupted by an *easterly wave*. When this happens, from one to three days of constant cloudiness follows before normal conditions return.

PRECIPITATION

From May through the summer and fall, brief but intense tropical showers are common. These may occur day or night. In December and January they become less common. From February through April they become rare.

Mid-August to mid-November is the hurricane season (see Chapter 12). In an average season about eight severe hurricanes form and pass this area. While these islands are seldom struck directly, they frequently get very heavy rains from the fringes of these storms.

VISIBILITY

There is a saying among sailors in the islands, "If you think you see fog—clean your glasses!" Fog is unknown here. Visibility is generally excellent. A typical tropical shower can reduce visibility to zero, but it will pass very quickly.

In some areas, particularly the south coast of Puerto Rico, industrial pollution has increased greatly in recent years. Industrial smoke may reduce visibility when the trade winds are light in such places.

HAWAIIAN ISLANDS

The weather in the Hawaiian Islands is very similar to that of Puerto Rico and the Virgin Islands. Both have a warm, humid climate, and both bask in the tropical sun gently cooled by steady trade winds. Hawaii is more mountainous with its high volcanoes, and, due to the long fetch of the Pacific, is washed by longer, higher ocean swells.

WINDS

The dominant wind is the constant NE trade wind. This wind is persistent year round except for occasional periods in the fall when they may weaken or die out temporarily.

In limited coastal areas on the west side of islands with high mountains to the east the trade winds are blocked. The result is SW winds with daily land / sea breezes.

CLOUD COVER

A scattering of fair weather cumulus clouds is typical of an average day. Completely clear days are rare. Complete overcasts are also unusual except during rain squalls, "kona weather" storms, or typhoons. "Kona" is Hawaiian for leeward. Kona storms come mostly from mid-fall through to April and bring heavy rain, strong winds, and thick cloudiness to areas that are normally on the lee sides of islands. During konas, normally safe leeward anchorages may become dangerous for small craft.

PRECIPITATION

Due to the mountainous terrain, rainfall varies considerably from place to place. Showers fall almost every day on the windward sides of islands and at higher elevations while the lower lee sides remain mostly dry except during "kona weather."

Hawaii lies between two major Pacific tropical storm areas. Since it is at the extreme edge of both, it seldom gets directly hit by storms from either one. Between May and November, the hurricane season in the east side of the North Pacific, the fringes of a hurricane sometimes hit the islands bringing wind, rain, and heavy seas.

VISIBILITY

Just as elsewhere in the trade wind belt, fog is virtually unknown along the Hawaiian coasts. Visibility is excellent except during tropical showers, kona weather, or the nearby passage of a tropical cyclone.

5
FOG

Nowhere are the coastal waters, inland waters, little lakes, or Great Lakes of the U.S. immune from fog. Depending on location and season of the year, fog varies from an occasional minor nuisance to a frequent and major hazard to navigation. In many places the same conditions that cause the formation of fog trap smoke and other air pollutants to produce "smog," a common cause of serious health problems.

Fog is comprised of the same type of minute water droplets that form clouds. Although the formation process is different, fog is actually the same thing as a stratus cloud except that it exists at surface level. Fog also forms for the same reason clouds do: a volume of air is cooled to the point where it can no longer hold all the water it contains in vapor form. Some of the vaporized water then precipitates as very tiny water droplets, small enough and light enough to remain suspended in the air.

DEW POINT

In order to understand how fog forms—or clouds, too, for that matter—we need to be clear on the meaning of *dew point* or *dew point temperature*. This is the temperature at which a particular parcel of air has become saturated. Air is considered saturated when it contains all the water it can possibly hold in vapor form, and has therefore reached a relative humidity of 100 percent.

When passing over oceans or other large bodies of water, the atmosphere picks up water by evaporation; that water changes state from liquid to vapor. The warmer the air, the more water vapor it can absorb. Conversely, the cooler the air, the less water it can continue to hold in vapor form. A volume of air steadily cooling will finally reach its dew point. Cooling below this temperature must result in condensation of part of its water vapor back into liquid water.

In Chapter 3, the use of the sling psychrometer was discussed in connection with determining relative humidity. By consulting the Dew Point Table (Appendix VI) the same instrument may be used to determine the dew point temperature as well.

When the air temperature and the dew point are well separated, and remain so, fog will not form. When air temperature and dew point temperature are moving closer, beware of fog—when the difference disappears fog will form. The warming of air increases the spread between air temperature and dew point which is called the *dew point spread.* Cooling decreases that spread.

By passing over water, air picks up moisture raising its dew point temperature toward the existing air temperature, and decreasing the dew point spread. As vapor condenses back to liquid form, the dew point temperature of the air drops. Fog may be formed either by air cooling until it reaches its dew point temperature, or by adding water vapor until the air temperature and dew point temperature become the same.

TYPES OF FOG AND FOG FORMATION

Air cools to a temperature below its point of saturation for any of several reasons. Consequently, there are several different types of fog. At this point you may well be thinking, "Fog is fog is fog! It all looks the same, it's all gray and I can't see through it—who cares about *different types?*" On the one hand you are absolutely right—it *does* all look pretty much the same. On the other hand the different types do matter. Through becoming familiar with the different types of fog, and what causes them, you'll know when to expect it, and can prepare to deal with it.

ADVECTION FOGS

Advection is the horizontal movement of air. Advection fog occurs when moist air that is comparatively warm moves horizontally over a colder surface. The lower layers of that air lose heat to the cold surface. The dew point spread closes, and water vapor starts to condense on salt, dust, or other particles, even before the air reaches its dew point temperature.

About 80 percent of all marine fog is advection fog, so it is by far the most common type encountered by sailors. Winds somewhere between 4 and 15 knots are needed for its formation. Usually, but not always, if the wind is stronger than 15 knots the fog lifts off the surface and becomes a low stratus cloud. I say not always because, as an example, I can recall once sitting for three days at anchor in Great South Bay off Long Island, New York stuck in a thick advection fog with a constant 18 to 20 knot wind the whole time! So at times the wind does fail to lift it.

Air in which advection fog has formed is air that has been cooled. Through cooling this air has become heavy and dense and thus does not tend to rise. Over land, when the winds are over 15 knots, the surface irregularities tend to mix the surface air through a comparatively deep layer often resulting in the formation of low stratus clouds rather than fog. Over the relatively smooth surface of the sea less turbulence develops, allowing advection fogs to form and persist at sea level in wind speeds as high as 30 knots.

The coast of California (Fig. 5-1) is one area particularly plagued by advection fogs. Along this coast the prevailing westerly winds produce an upwelling of cold water close to the coast. In summer these winds are northwesterly. This causes the coldest upwelling to occur in Northern California. The comparatively warmer, moist marine air passing over this cold water produces the summer fogs of Northern California.

In the winter the winds are mostly west to southwest. Warm subtropical marine air then strikes the cold water off Southern California often fogging in Los Angeles, San Diego, and other points on the U.S. coast, and well down into the Baja California peninsula of Mexico.

LAND BREEZE / SEA BREEZE FOG

The common coastal land breeze / sea breeze wind sequence was discussed in Chapter 3. Summer in eastern New England frequently offers particularly favorable conditions for this sequence to produce land breeze / sea breeze fogs, another type of advection fog.

In this case the summer-heated air is cooled as it moves out over the sea. When winds are light dense surface fog may develop over the water. Light afternoon sea breezes may bring this fog in over land, only to recede again with the night land breeze. In some cases a weak movement of air from offshore may prolong the fog a day or two.

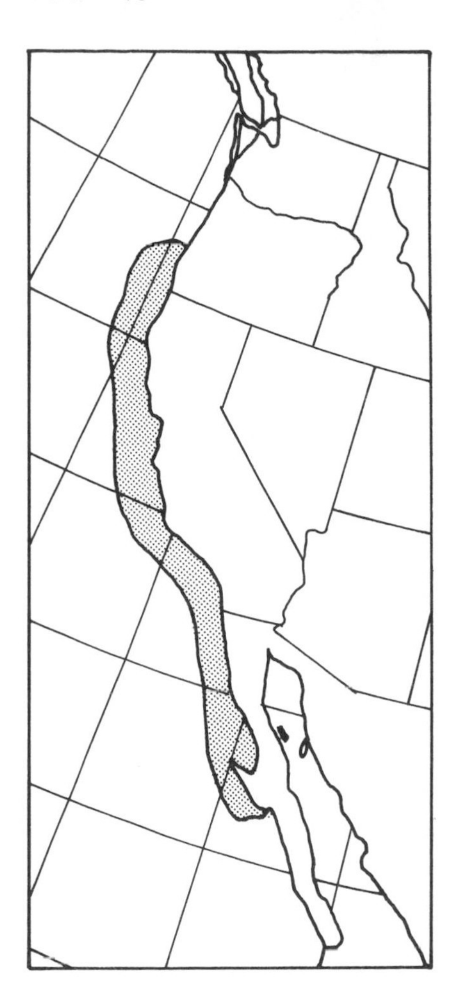

Figure 5-1. Areas of advection fogs along California coast.

SEA FOG

The notorious Grand Banks fogs of the North Atlantic are an example of advection sea fog which can occur anywhere over cold coastal currents. Such currents exist along the northern part of the U.S. Atlantic Coast as well as much of the Pacific Coast as mentioned earlier.

In winter, considerable cold, fresh river water flows into the Gulf of Mexico. Being fresh water, it is less dense than the warmer salt water of the Gulf, and tends to remain at the surface. Here it cools the bottom layer of moist surface air, often producing advection sea fogs over the Gulf (Fig. 5-2).

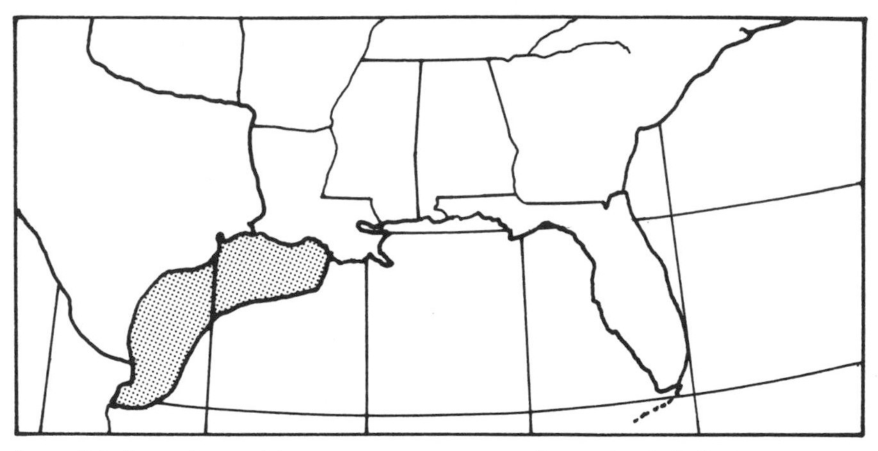

Figure 5-2. Area where cold river waters cause winter fogs in the Gulf of Mexico.

TROPICAL-AIR FOG

While this is another form of advection fog, it is not caused by warm air moving over a colder surface. In this case, warm moist tropical air is gradually cooled by moving northward from lower to higher latitudes. Very widespread fogs are produced in this way over both land and sea. Over the continental U.S., some of the most extensive fogs that occur are of this type. However, tropic-air fogs are more common over sea than land because the irregularities of the land cause so much more turbulence in the surface layer of air. This turbulence deepens the surface layer and lifts the fog to form low stratus clouds.

RADIATION FOG—GROUND FOG

Radiation fog occurs over land at night, when the sky is clear, allowing unobstructed radiation, and when winds are very light. In the evening, as has been mentioned earlier, the land cools considerably and rapidly, while the sea cools very little. Over land, at times when the relative humidity is already high, a fairly modest drop in air temperature may be enough to reach the dew point. A very light breeze then completes the conditions favorable for the formation of radiation or ground fog.

The breeze must be very light—2 to 4 knots—so as to produce a thin layer, only 50- to 100-feet thick, of gentle surface turbulence. This forms a low band of cool air in which a solid fog can form. In absolutely calm air, the cool layer directly at ground level is too thin for fog to form, and if there is too much wind, say 10 to 15 knots, the turbulent mixing layer becomes too thick hindering, or entirely blocking fog formation. Cool air drains downhill so radiation or ground fogs are thickest in low lying valley bottoms. Surrounding hills will generally stand above the ground fog.

Radiation fog does not form over water because its surface temperature changes very little between day and night. However, it is of concern to the sailor because it often obscures beacons and landmarks ashore making night navigation difficult. Radiation fogs generally dissipate two to three hours after sunup. Fall and winter are the most likely seasons for radiation fogs, particularly during periods of high pressure, very clear weather, and light winds.

FRONTAL FOGS

Cloudy, rainy, and generally dreary weather conditions occur along the dividing lines, called *fronts* between large air masses of differing characteristics. The various kinds of air masses and frontal conditions are discussed in detail in Chapter 10. One of the dreary conditions that often accompanies the passage of a weather front is fog.

When a mass of warm air meets a mass of cooler air, the warm air rises over the cool because it is lighter and less dense (see Fig. 10-7). As it rises it cools. Continuing to rise and cool, it reaches its dew point, water vapor condenses, and clouds form. As it goes on rising and cooling, further condensation produces rain which falls through the warm air layer and down into the colder air below (Fig. 5-3). The falling warm rain evaporates in the cool layer which, as we have seen, raises the dew point of that layer. When the winds in advance of the front are very light fog may form in advance of the passage of the warm front. This type of fog is called *prefrontal fog* or *warm front fog*.

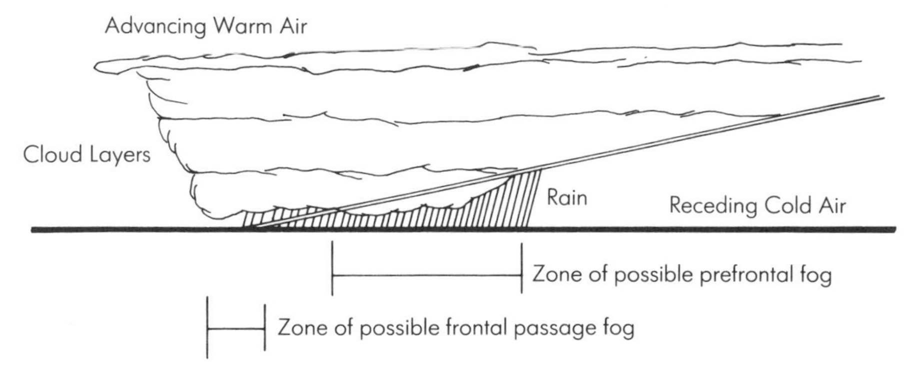

Figure 5-3. Prefrontal or frontal fogs may occur when winds are very light at the time when two air masses of differing characteristics are passing through an area.

Another type of frontal fog is termed *frontal passage fog*. This sometimes occurs along with the passage of a front, rather than in advance of it. Again winds must be very light. When both warm and cool air masses are close to saturation, the mixing of the two along the frontal zone may briefly produce fog. Also, cooling of the ground by evaporation of rainwater that fell during the prefrontal rains may both cool and saturate a narrow band of air along the front, causing fog for a short time.

Postfrontal fog or *cold front fog* is far less common than the prefrontal type, but it does occasionally occur. It too is caused by evaporation of falling rain. Since the rain band associated with a cold front is much narrower than that accompanying a warm front, postfrontal fogs are much less widespread. They are most likely to occur when the cold mass is stable (see Figs. 10-14a and c). If it is not stable fog will not form (Figs. 10-14b and d).

AREAS OF FOGS IN THE UNITED STATES

Since the conditions that cause fogs differ, and the climate and terrain vary so greatly over the Continental U.S. the frequency of fogs also varies considerably from place to place. Figure 5-4 shows the country divided roughly into 12 major areas based on the frequency of the occurrence of fogs.

CALIFORNIA COAST *(Fig. 5-4 Area 1)*

While the State has the reputation of "Sunny California," its coastal areas also have the highest incidence of fog in the country with the sole exception of Nantucket Island, Massachusetts. The California coastal fogs are primarily advection fogs, as discussed earlier.

The upwelling of cold waters along the coast is constant. However, the source of the warm, moist air that passes over it to produce advection fogs changes seasonally along with the winds that carry that air. The famous San Francisco fogs that are most frequent in July and August develops from moist air flowing from the north side of the Pacific anticyclone (see Fig. 10-5). Los Angeles, San Diego and Southern California are fogged in most often in December and January by moist maritime tropical air flowing up from the southwest (see Fig. 10-3).

Due to the velocity of the winds, these fogs often move up just a very few hundred feet becoming actually very low stratus clouds. This was the reason for the strange case in which the lighthouse on

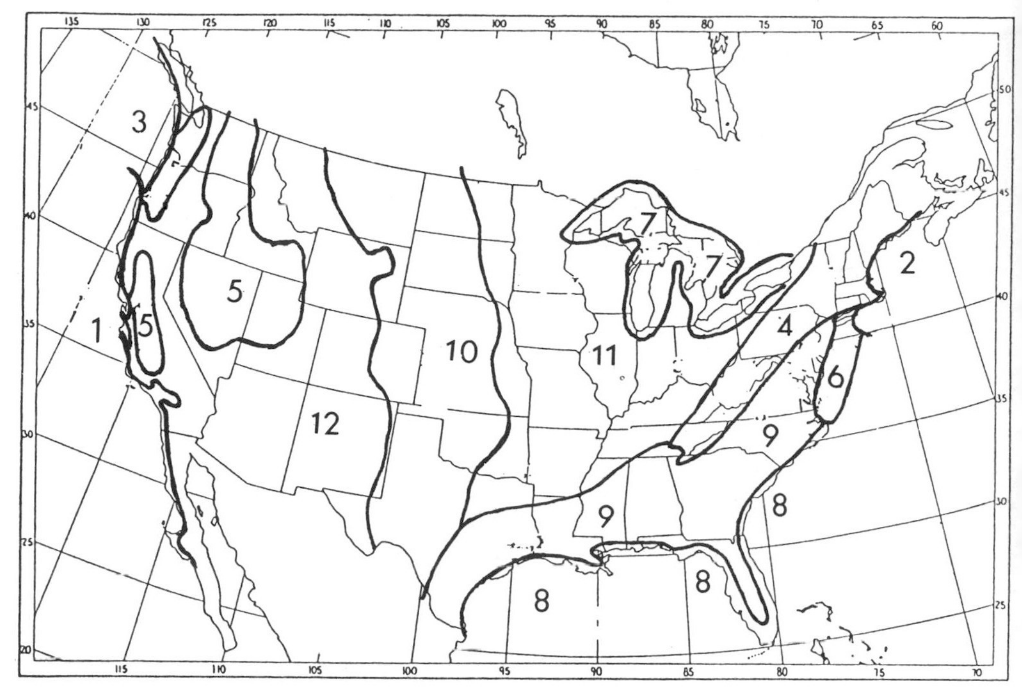

Figure 5-4. Areas of United States in order of frequency of fogs. 1. California Coast; 2. New England Coast; 3. Northern Pacific Coast; 4. Appalachian Valleys; 5. Pacific Coast inland valleys; 6. Middle Atlantic Coast. 7. Great Lakes; 8. Southern Atlantic and Gulf coasts. 9. Inland Atlantic and Gulf plains. 10. Great Plains; 11. Central river valleys—Ohio, Missouri, Upper Mississippi; 12. Western plains and deserts. Courtesy: Beyers, *General Meteorology*.

Point Loma, at the entrance to San Diego harbor, was moved down from the top of the Point almost to the bottom. All too frequently the top of the Point where the light stood 450 feet above sea level is(seeshrouded in fog, hiding the light. The new light, only 88 feet above sea level, is at an altitude that is usually clearer.

NEW ENGLAND COAST (Fig. 5-4 Area 2)

Coastal New England fogs are most frequent during the warmer months of the year. Winds are predominantly in the quadrant between south and west bringing warm humid air with them. As this air moves over the cool waters lying between the coast and the offshore Gulf Stream, land breeze / sea breeze fogs are formed.

In winter, tropical-air fogs may form in air masses that occasionally work their way up the east coast. Also, pre-frontal fogs often form in advance of the warm fronts of winter cyclonic storm systems. In fact, the prefrontal fogs of New England are probably the best developed examples in the world of this type of fog.

NORTHERN PACIFIC COAST (Fig. 5-4 Area 3)

Along the North Pacific Coast fogs are again most frequent in summer, peaking in August. However, the types and causes are different from those that occur further south in California. Tropical-air fogs and sea fogs are the types typical of this area.

The exceptionally cold coastal current of California is absent, but here coastal waters are significantly colder than the open ocean. Light summer winds bring tropical air from the Pacific anticyclone across the cool coastal waters forming a combination of tropic-air and sea fog. Since winds are light the fogs, once formed, remain at surface level rather than rising as the California fogs often do.

Winter prefrontal fogs, such as those that occur in the Northeast, are not found in the Northwest due to the stronger winter winds.

APPALACHIAN VALLEYS (Fig. 5-4 Area 4)

The principal fogs of this area are radiation fogs formed by radiation cooling and air drainage down into sheltered valleys and lowlands at night. The air in these parts is most humid in July and August, and quite humid in June and September as well. With the shortest nights of the year in June and July, the radiation cooling in those months is not as great as in August and September when the nights grow longer. Consequently, August and September are the months when conditions are most favorable for formation of radiation fogs.

Prefrontal warm front fogs are the usual winter fogs of this region. These are similar to the New England winter fogs of the same type. While prefrontal fogs occur most frequently in winter they may occur at any season of the year.

PACIFIC COAST INLAND VALLEYS (Fig. 5-4 Area 5)

Fogs in this area are largely radiation fogs occurring during the cold season from late fall through winter to early spring. Radiation fogs usually dissipate shortly after sunrise. However, in this region they tend to be more persistent than normal, sometimes lasting for several days. This is not a major boating area; however, you may want to remember this in case you take your boat to fish the lakes here in winter.

MIDDLE ATLANTIC COAST (Fig. 5-4 Area 6)

The waters from Cape Cod to Cape Hatteras are coldest during late winter. In addition, at this season a number of large rivers discharge fresh water that is near freezing. The fresh water being lighter than salt now stays at the surface, making the surface waters exceptionally cold. The temperature contrast with the warm offshore Gulf Stream now sets up ideal conditions for the formation of advection fogs. These winter advection fogs are purely a coastal phenomenon disappearing a few miles inland.

In summer, the typical land breeze / sea breeze fogs sometimes occur, but more often, the warm weather fogs are sea or tropic-air fogs. Also, as along the coast further north, prefrontal fogs may occur at any season.

GREAT LAKES *(Fig. 5-4 Area 7)*

Like the coastal oceans, the waters of the Great Lakes warm more slowly in spring, and cool more slowly in fall than the surrounding land. In spring, as the land becomes warmer than the Lakes, conditions are good for land / sea breeze type fogs. In fall, when the Lakes stay warm as the land cools, advection-radiation fogs occur. In winter, offshore steam fogs are encountered. These are likely to be far more of a problem for commercial shipping than for the pleasure boat operator who ought to be snugly settled at home during this season.

SOUTHERN ATLANTIC AND GULF COASTS *(Fig. 5-4 Area 8)*

While conditions along these coasts are rather similar to the Middle Atlantic Coast, fogs here are less frequent. In late winter and early spring, cold river discharge produces inshore waters colder than those offshore. As warm surface air crosses from the Gulf Stream or the Gulf of Mexico, fogs may form. Since the temperature contrasts are not as great as along the Middle Atlantic, fogs are not as frequent. On the peninsula of Florida, fogs are rare.

In this area, fogs are most frequent in December and January. They usually disappear entirely in April or May, and do not reappear until late fall.

INLAND ATLANTIC AND GULF PLAINS *(Fig. 5-4 Area 9)*

The occasional fogs that occur in this area are largely radiation types. Sporadically, tropic-air fogs appear in spring or early summer. Also from time to time in winter, prefrontal fogs may form.

While the majority of the fogs in this region are radiation type appearing in late summer and fall, they usually dissipate shortly after sunrise causing little inconvenience. The winter fogs are less frequent, but when they do occur they are more persistent.

***GREAT PLAINS** (Fig. 5-4 Area 10)*
The fogs in this region are largely *upslope* fogs. This is a purely continental (non-marine) fog type occurring only over large land areas. Here the prairies gradually rise from sea level to an altitude of between 5,000 and 6,000 feet. Air cools as it rises and this cooling lowers its dew point. As we have seen, cool air cannot hold as much water in vapor form as warm air. So as cooling continues eventually the dew point temperature is reached and fog forms.

The fogs in this general region thus occur most frequently in the areas of highest altitude. Rainfall is also an important factor in the formation of fogs here. Evaporation after a cold front passage will often complete the saturation of air that has cooled nearly to the dew point after being forced upward over many miles of rising ground. Since surface cooling is a minor factor in these fogs, they can develop (like some advection fogs) in fairly strong winds. Dense surface fogs have been recorded at places such as Cheyenne and Amarillo in winds between 20 and 30 mph.

***CENTRAL RIVER VALLEYS** (Fig. 5-4 Area 11)*
Within this region there are great differences in fog frequency between the bottom lands of the river valleys, and higher surrounding ground. Highly localized radiation ground fogs form at night in the low lands only. These fogs dissipate during the day.

Both prefrontal warm front fogs, and postfrontal cold front fogs also occur in this area. Fog frequency peaks here in January and almost disappears from May to August.

***WESTERN PLAINS AND DESERTS** (Fig. 5-4 Area 12)*
Fog in this region is very infrequent. As in the Central River Valleys, radiation fog occasionally forms in low lands at night dissipating rapidly in the morning. Occasional frontal fogs also occur as major weather systems move through the area.

PREDICTING FOG

For the sailor, fog prediction is basically a local matter. In home waters, you are likely to be familiar with the seasons and local temperature and weather conditions that have accompanied previous occurrences of fog. When in unfamiliar areas seek out local knowledge.

ADVECTION FOG

Advection fog is difficult to forecast from the boat. Because of the way it forms, you need, in addition to the sea surface temperature, a knowledge of future wind direction and velocity as well as air temperature and humidity. Weather Service often has this information, and, when they do, it will appear in their broadcasts. Such broadcasts, you will recall, cover comparatively large areas. When they predict a general fog, you're probably included, so prepare. When they predict scattered, or patchy fog you may or may not get it, but be ready anyway. In general, whenever warm, moist air is blowing over cold water advection fog is possible.

Remember, advection fog is going to form in air that has not yet reached your location. However, in daylight hours it can be seen coming for a considerable distance over water.

Also, watch the general sky condition. Advection fog can come in under a completely clear sky, or it can come in under progressively lowering stratus clouds (see Chapter 9). It is unlikely in the unstable air accompanying cumulus type clouds.

RADIATION FOG

Radiation fog is going to form at night, under a clear sky, in air that is relatively humid, and with a very light wind. If it is cloudy, or there is no wind at all, or there is a strong wind, or the humidity is very low radiation fog will not develop.

When conditions are favorable for radiation fog, a sling psychrometer will help to determine not only whether fog is likely, but also the time it will probably appear. To do this, take a reading at sundown, and then once every hour for several hours during the evening.

Suppose that at 6:00 PM (1800) the skies are clear, wind is down to 4 knots and your psychrometer gives you an air temperature of 74°F and a dew point of 55°F. You have heard a broadcast prediction of 50°F as the low for tonight. Your hourly readings for the next few hours are:

TIME	AIR TEMP.	DEW POINT
6:00	74°F	55°F
7:00	71°F	55°F
8:00	68°F	55°F
9:00	65°F	55°F
10:00	62°F	55°F

The air temperature is approaching the dew point by 3°F per hour. At this rate fog is possible sometime after 1:00 AM. In reality the dew point spread seldom narrows this evenly. The spread is more likely to converge more slowly as the night progresses. Thus in this case the fog might not appear until 2:00 AM or later.

In general, radiation fogs are most common in late summer and early fall when the daytime air is still warm and humid, but the nights are getting longer leaving more time for radiation cooling. By winter, when the nights are longest, the cold has made the air too dry, and the nights often too windy for radiation fogs.

PREFRONTAL FOG

This type of fog is often part of a large dreary rainy area preceding a slow moving warm front. Once it has developed, it will probably persist until the front has passed and the rains stop. Any time the passing of a warm front has been predicted, or you see the typical cloud progression that signals the coming of such a frontal system (see Chapter 10) you should expect a prefrontal fog to come with it.

NAVIGATING A SMALL BOAT IN FOG

Of course the first major point to remember about piloting a small boat in fog is: if at all possible—DON'T DO IT. Stay in port, or stay at anchor until it lifts.

With that clearly understood, let us now assume that it has become absolutely necessary to move the boat through a fog. You will need:

- An up-to-date chart.
- A reliable compass.
- A reliable watch with sweep second hand or stop watch capability.
- An accurate knotmeter or speed curve for your boat at various RPMs.
- An adequate horn (to be used in accordance with the Pilot Rules).

If any of these items are missing, think again. Do you *really* have to get under way? If you are already under way and are caught in a fog, can you possibly drop anchor and stay there until the fog lifts?

The worst case scenario would be one where you are caught under way in fog in conditions such that it is impossible to anchor—the water is too deep, or wind, tide, or currents make anchoring unsafe. This can happen to any of us—don't panic! You are now going to work your way slowly, carefully, and deliberately to safety.

When you first saw the fog coming you got a fix on your position, and plotted it on the chart, so you know with reasonable accuracy where you now are. Forget the direct course to your destination, from your present position you are going to plot a course to the nearest buoy. Then you'll work your way by compass courses from buoy to buoy making the runs as short as possible. This way you can make frequent checks on your progress. If you try to make long runs, the combination of steering errors plus wind and current drift may take you well off your intended track.

Post lookouts bow and stern, and sound proper fog signals. Proceed very slowly, and stop engines from time to time to listen for nearby boats, fog signals from beacons or other aids to navigation, or any other sounds you can pick up.

The apparent direction of sounds in fog can be very deceptive. Consequently, do not abandon your compass course because of the apparent direction of a sound. Check your compass carefully in clear weather, and then in fog *trust* it. Steer in a fog only by compass, not by hunch. Once its deviations are known, a compass is reliable, your hunches are not.

On the chart, measure the distances from buoy to buoy. Then set a fixed speed at which you will run. This will allow you to calculate the time it should take you to get from one to another. The formula for this is:

$$60 \times D = S \times T, \text{ or } 60 \times \text{Distance} = \text{Speed} \times \text{Time}$$

For example, the distance between two buoys measures out to .6 nautical miles. You plan to proceed dead slow at 3 knots. You should reach the next buoy in 12 minutes:

$$60 \times .6 = 36$$
$$36 \div 3 = 12$$

As long as the buoy to buoy runs work out properly, continue making them. If one of them fails to work out so that at the expected time nothing is found, *stop*. Check the calculation for mathematical error. Mistakes under pressure are not uncommon. If the math was right, make a 90° turn and start a square search pattern (Fig. 5-5). Continue the search until you find the buoy, and then pick up where you left off.

If you haven't found the buoy in a half to three-quarters of an hour, stop again and anchor, or heave to, and wait for the fog to lift.

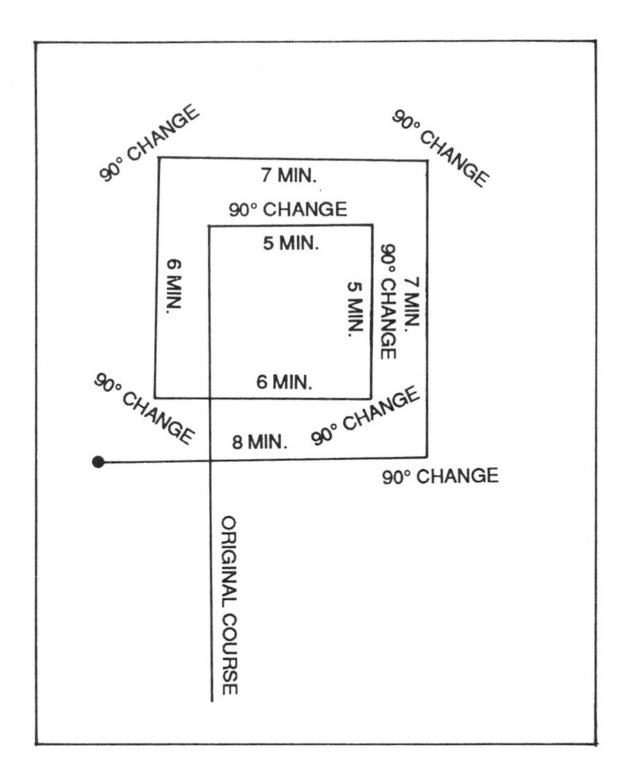

Figure 5-5. Square search pattern. From: Markell, *Coastal Navigation for the Small-Boat Sailor.*

At that point, shut down engines to save fuel, and shut off everything electrical except navigation lights to conserve battery power for engine starting or emergency radio transmission, should really serious trouble develop.

Remember, while many situations can arise at sea in which the boat and the people in her are truly in danger, merely being lost in a fog is *not* one of them. If you have someone aboard who is severely ill or injured, or the boat is in sinking condition you have legitimate reason to call Coast Guard for assistance. However, if you are just rather cold, hungry, and uncomfortable, and suddenly fresh out of beer and cigarettes as well, too bad! None of this constitutes adequate justification to call out people and equipment who may be needed elsewhere to help someone who has a *really* serious problem.

6

WEATHER AND THE OCEANS

Since approximately four-fifths of our planet is covered by oceans, the influence of oceans on our atmosphere and weather and influence of the atmosphere on the oceans, is immense. The causes of the many dramatic effects of the two on each other can be traced to the absorption, transfer, and redistribution of the heat received from incoming solar radiation.

Some of what we observe results directly from the effects of solar heat on air and water. The origin of the winds, however, is indirect. The winds, and their many consequences, result from differences in atmospheric pressure that, in turn, are caused by the uneven heating of the atmosphere.

EFFECTS OF THE OCEANS ON THE ATMOSPHERE

The oceans absorb vast amounts of solar heat. Much of the heat remains in the first few feet below the surface. However, due to wave and current action, part of it mixes into considerably deeper layers. Because of this mixing, all large bodies of water act as huge "heat sinks." In Chapter 8 this is noted along with the explanation of the important distinction between *heat* and *temperature*.

HEAT ABSORPTION AND TRANSFER

Although the oceans absorb huge amounts of heat, it is distributed through such large masses of water that the temperature changes very slowly. Conversely, air temperatures tend to fluctuate far more rapidly and over a much wider range. Due to the comparative stability of its temperature, a body of water has the effect of reducing the temperature variations in the air above it. Water tends to cool air that is warmer than it is, and warm air that is cooler by directly conducting heat to or from the atmosphere.

MARITIME AIR MASS FORMATION

One major direct influence of oceans on the atmosphere is the creation of maritime air masses (Chapter 10). When a large volume of air lingers over a portion of the sea for a number of days, the temperature of the entire mass moves toward that of the water while picking up humidity through surface evaporation. These large air masses become primary factors in the great traveling weather systems that move about our planet (see Chapters 10 and 11).

SEA AND ADVECTION FOG

In the above-mentioned instances, when the two substances, air and water, have different temperatures and are in direct contact, heat flows from the warmer one to the cooler. Thus when warm, humid air passes over cold ocean water the bottom layers of that air are cooled, often to their dew point, resulting in advection fogs as we saw in the last chapter. These fogs are another direct and highly visible effect of the ocean on the atmosphere.

CLIMATIC EFFECT OF OCEAN CURRENTS

The moderating effects of oceans and ocean currents on the weather and climate of adjacent land areas are not as immediately perceptible as fog, but they too can be extremely drastic. For example, along the coast in summer in Southern California temperatures seldom exceed the mid-eighties while five to ten miles inland temperatures well over one hundred degrees are not uncommon. The cool California Current close to the coast keeps the coastal temperatures down in summer.

In winter, in the same area, water temperatures drop only to the low fifties. In consequence frost and snow do not occur along the coast. However, a short distance inland snow is found in the nearby mountains as low as 4,000 feet, and nighttime frosts occasionally penetrate to within a couple miles of the coast. The temperature of the sea varies with the seasons much less than that of the air so the sea cools the air in summer, and warms it in winter.

THE WINDS

The immediate cause of wind is a difference in atmospheric pressure. This pressure difference, in turn, is caused by differences in air temperatures. The oceans get into this act, as we have just seen, by altering air temperatures.

The local land breeze / sea breeze sequence mentioned in Chapter 3, and the fog that sometimes accompanies it, is a good case of a localized effect of the sea (combined with that of the land) on atmospheric temperature and pressure. During the day, air over the water is cooler than air over the land. Cool air is denser than warm air, and therefore has a slightly higher pressure. Air moves from high pressure to low, hence a breeze blows from sea to land.

At night opposite conditions take over. The land cools rapidly and cools the air above it. Air over the water now becomes warmer, less dense, and lower in pressure than over the land. This causes the breeze to blow from land to sea.

TROPICAL CYCLONES

Another important example of the effect of oceans on the atmosphere is the formation of tropical cyclones (see Chapter 12). These are variously known as hurricanes, typhoons, "baguios," "williwaws," or "chubascos." These intense *warm core* storms originate and behave totally differently from temperate zone *cold core* cyclonic storms (Chapter 11). They are much smaller in area, much more violent, and their movements are far more erratic.

They form only over warm oceans. The surface water temperature must be above about 80°F in order to start and maintain the strong vertical air circulation that builds the raging winds associated with these storms.

EFFECTS OF THE ATMOSPHERE ON THE OCEANS

We have seen that the oceans affect the atmosphere by absorbing heat from it, or giving up heat to it. These heat transfers result in atmospheric temperature and pressure changes. Pressure differences, as we have seen, cause winds. The oceans also add vast amounts of water vapor to the atmosphere through evaporation.

In return, the atmosphere has far-reaching effects on the oceans as its winds stir up waves, swells, and currents. All waves except the occasional seismic waves, and nearly all surface currents except the daily tidal currents or river discharge currents are originally wind generated.

WINDS AND WAVE FORMATION

Anyone who has ever sat becalmed in a sailboat, or any other kind of boat for that matter, knows that a very slight breeze of only a couple knots is enough to produce surface ripples. These "cats paws" form immediately with the breeze, and die when it dies.

When that light breeze becomes a steady wind, regular waves form that will persist after the wind dies down. The size of the waves thus formed depends on several factors:

1. Wind speed
2. Time the wind has blown
3. Distance (fetch) over which it has blown

The size of a wave is determined by its height, its length (Fig. 6-1), and its period. Wave period is the time in seconds required for one wave to pass crest to crest by a fixed point.

No matter how long the wind blows, nor over how great a fetch, the maximum size of the waves that will develop is limited by the wind speed. As the wind speed increases maximum wave height increases with it up to about 50 feet. Waves above this height are occasionally recorded, but they are uncommon. There is a mathematical explanation for this but it is not important for our purposes.

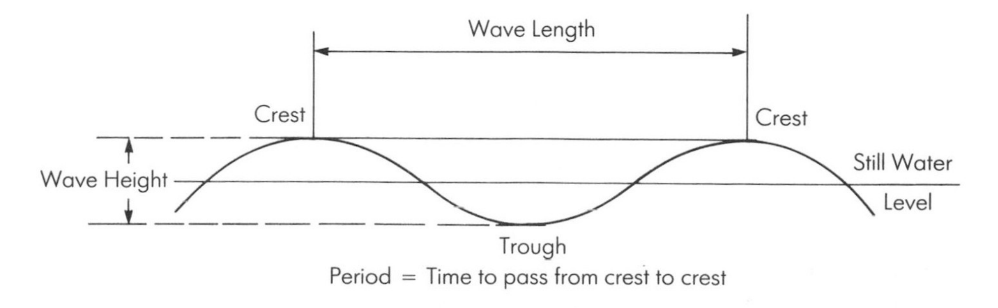

Figure 6-1. Characteristics of an ocean wave.

The time during which a wind acts, as well as its speed, is a major factor in determining wave height. Following is an approximation of wave heights that can be expected from several wind speeds blowing for different lengths of time:

WIND SPEED (in Knots)	TIME BLOWING (in Hours) 5	10	15	20	30	40
10	2	2	2	2	2	2
15	4	4	5	5	5	5
20	5	7	8	8	9	9
30	9	13	16	17	18	19
40	14	21	25	28	31	33
50	19	29	36	40	45	48

Not many small-boat sailors are going to be at sea for 30 to 40 hours in winds of 30 to 40 knots, but any of us might well be out for a while in 15 to 20 knot winds. If so we might find ourselves dealing with 8- or 9-foot seas. In open ocean in a well-found boat this might be uncomfortable, but by no means dangerous. When such seas approach shoal water or an inlet with opposing tidal or river current, conditions may quickly become extremely dangerous.

THE BEAUFORT WIND SCALE

In the early 1800s, Admiral Beaufort, of the Royal Navy, worked out a scale of wind velocities (Fig. 6-2) with accompanying descriptions of the sea conditions that these velocities produce. This was well before the invention of our current anemometers and was intended to serve as a guide for sailing ship officers. With it, they used the observed sea condition to indicate when it was time to shorten sail, or conversely when it was safe to shake out a reef.

The sailing ships of the Royal Navy for whom this scale was devised are long gone. However, the scale is still a very useful guide for the modern mariner as to when to shorten sail or reduce engine speed.

WAVES IN SHOAL WATER

As waves move from deep water into shoal water, or approach the shore, their shape alters considerably. When the water depth shelves up to about half the wave length, the wave is said to "feel the bottom." Its height increases while its length decreases. If the height-to-length ratio gets down to somewhere between 1:7 and 1:10 the wave becomes very steep, unstable, and likely to break (Fig. 6-3).

All too often when approaching an inlet or channel from sea, you are unable to tell how bad conditions are in the channel until you are

THE BEAUFORT WIND SCALE

BEAUFORT NUMBER	KNOTS	MILES PER HOUR	DESCRIPTION	EFFECT AT SEA	WIND SYMBOLS ON WEATHER MAPS
0	0–0.9	0–0.9	Calm	Sea like a mirror.	Calm
1	1–3	1–3	Light air	Scale-like ripples form, but without foam crests.	Almost Calm
2	4–6	4–7	Light breeze	Small wavelets, short but more pronounced. Crests have a glassy appearance and do not break.	5 Knots
3	7–10	8–12	Gentle breeze	Large wavelets. Crests begin to break. Foam has glassy appearance. Perhaps scattered white horses.	10 Knots
4	11–16	13–18	Moderate breeze	Small waves, becoming longer. Fairly frequent white horses.	15 Knots
5	17–21	19–24	Fresh breeze	Moderate waves, taking a more pronounced long form. Many white horses are formed. Chance of some spray.	20 Knots
6	22–27	25–31	Strong breeze	Large waves begin to form. White foam crests are more extensive everywhere. Some spray.	25 Knots
7	28–33	32–38	Moderate gale	Sea heaps up and white foam from breaking waves begins to be blown in streaks along the direction of the wind. Spindrift begins.	30 Knots
8	34–40	39–46	Fresh gale	Moderately high waves of greater length. Edges of crests break into spindrift. Foam is blown in well-marked streaks along the direction of the wind.	35 Knots
9	41–47	47–54	Strong gale	High waves. Dense streaks of foam along the direction of the wind. Sea begins to roll. Spray may affect visibility.	45 Knots
10	48–55	55–63	Whole gale and/or Storm	Very high waves with long overhanging crests. The resulting foam in great patches is blown in dense white streaks along the direction of the wind. On the whole, the surface of the sea takes a white appearance. The rolling of the sea becomes heavy and shocklike. Visibility is affected.	50 Knots
11	56–63	64–73	Storm and/or Violent Storm	Exceptionally high waves. Small- and medium-sized vessels might for a long time be lost to view behind the waves. The sea is completely covered with long white patches of foam lying along the direction of the wind. Everywhere, the edges of the wave crests are blown into froth. Visibility seriously affected.	60 Knots
12	64 or higher	74 or higher	Hurricane & Typhoon	The air is filled with foam and spray. Sea is completely white with driving spray. Visibility is very seriously affected.	75 Knots

From: Weather for the Mariner—Admiral Kotsch—Naval Institute Press

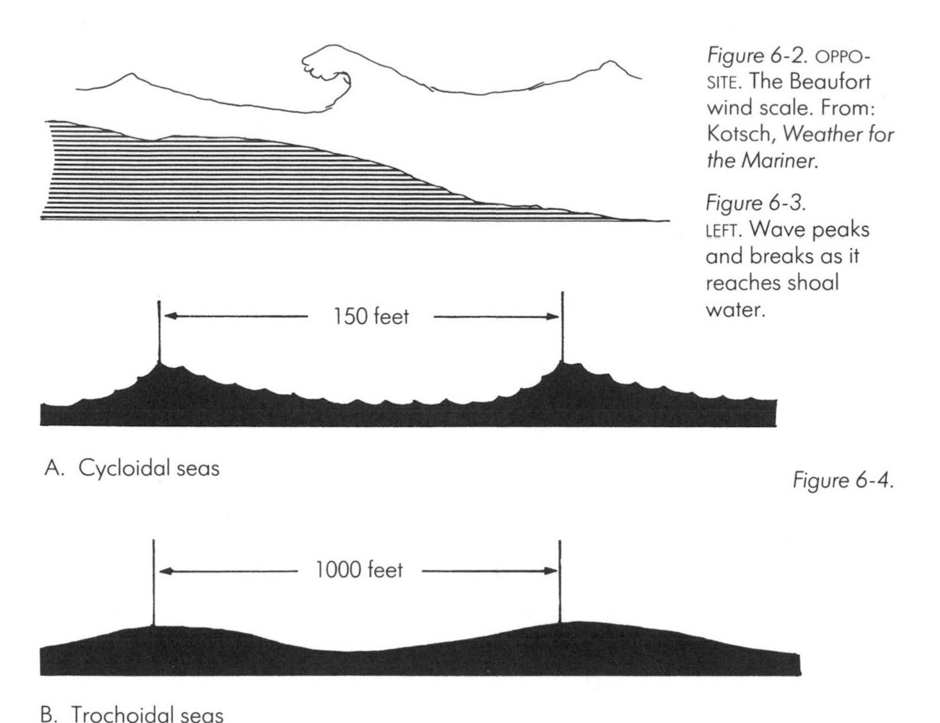

Figure 6-2. OPPOSITE. The Beaufort wind scale. From: Kotsch, *Weather for the Mariner.*

Figure 6-3. LEFT. Wave peaks and breaks as it reaches shoal water.

Figure 6-4.

already in trouble. Running a breaking inlet is a job for only the most skilled boat handlers, and most of them have the wit to avoid it if possible. If there is any doubt at all about conditions in the channel, stay well out in open water until the seas die down. It may be uncomfortable to lie outside, but trying to enter could make matters a lot worse.

SHALLOW WATER WAVES

Waves that form in shallow water are quite different from those that form in deep water. Even when wind speed is considerable, the water depth, or rather lack of it, limits both wave height and length. Much shorter and steeper cycloidal waves (Fig. 6-4) develop in shallow water than a wind of the same velocity would cause in deep water where trochoidal waves are formed. Also, they start to break sooner. Thus, even moderate winds can quickly produce very nasty conditions in large shallow lakes or bays.

WAVES AND CURRENTS

An opposing current will decrease wave length and increase height without altering the wave period. A trochoidal wave becomes cycloidal. Decreasing the height-to-length ratio is again moving the wave toward instability and the likelihood of breaking (see Fig. 6-3). A quite moderate sea entering an inlet or channel against an ebb current may thus become very dangerous until the current slacks.

Again, from seaward it may be impossible to tell how bad conditions are in the channel. When entering a channel with which you are not familiar, consult your Current Table so as to enter with slack water, or, better yet, with a flood or following current. A following current lengthens the waves and reduces height. This increases the height-to-length ratio, reduces wave steepness, and gives you a smooth ride in.

SWELLS

Once waves have been formed, the wind that formed them may die down, or the waves may pass out beyond the influence of the wind system that started them. With the wind no longer acting on them, waves begin to lose their energy. Small wind wavelets gradually disappear until all that is left are the long, smooth, round-crested swells, called trochoidal waves (see Fig. 6-4). As long as they remain in deep water these swells roll on in their original direction regardless of what the wind is doing in the area they have now reached.

For this reason swells encountered at sea may be moving at any angle to the existing wind. It is commonplace to sail through wind driven waves from one direction that are being crossed by long swells from another. Also, swells may come from more than one direction at any given time. Because of wave refraction and reflection it is common when sailing for example in the Caribbean islands or other island chains, to find swell trains coming in from as many as three different directions, none of which aligns with the present wind.

Acute sensitivity to these various crossing swells is one of the secrets of the amazing navigational skill of the Polynesians. They crossed thousands of miles of open ocean with no charts, no watches, no sextants, no navigational tables, and no compasses to unerringly find small islands by reading only the stars, the winds, the weather, and the wave patterns of the sea.

Because of the change in wave shape as a swell feels the ground, keep a small outboard or sailboat well outside the breaker line in even a very light ground swell. A long, low two-foot ground swell can build up to a very impressive four or five-foot breaker that doesn't look big from the sea side until you are in it. By then it is often too late.

In the movies, on TV, or perhaps even at the waterfront you may have seen boats taken in to a beach, or out from it through the surf.

When done by a trained Coast Guard surf crew, or a crew of working professional fishermen it looks easy. Don't be fooled—it's not. Don't try it unless you *really* know what you're doing.

REFRACTION OF SWELLS

As an evenly spaced train of waves or swells approaches a curved shoreline, the waves are bent, or *refracted.* This is caused by the contours of the sea bottom. As the water shoals, the wave length decreases while its height increases, bending the straight line of the wave back as it advances (Fig. 6-5). This produces a curved progressive breaker line along the shore, and often a current parallel to the shore that can become strong enough to be very dangerous to swimmers.

Wave refraction around a small island can also produce some very interesting crossing swell conditions. In the immediate lee of the island (the lee side relative to the swell) the sea becomes completely calm except for wavelets stirred up by today's wind. However, as you move offshore on the lee side you begin to feel refracted swells first from one direction, and then from two! The original swell has been bent in different directions by the two ends of the island, and you are feeling both of them.

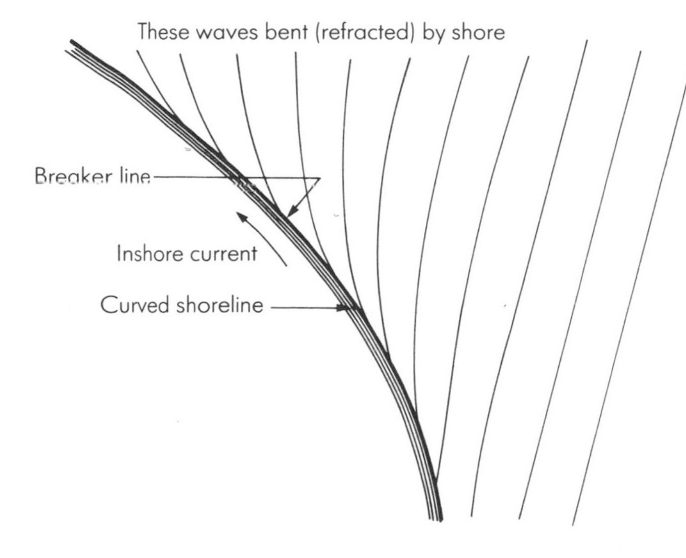

Figure 6-5. As waves reach shallow water, wave length decreases, wave height increases. This bends the shallow water end of the wave. The resulting curved breaker line causes an inshore current that can be dangerous to swimmers.

In the process of refraction some of the energy of the original swell has been used up. The two refracted swells are each smaller than the wave train they came from. However, if the original waves were fairly large, then the two refracted swells on the lee side can produce quite boisterous, and sometimes dangerous conditions.

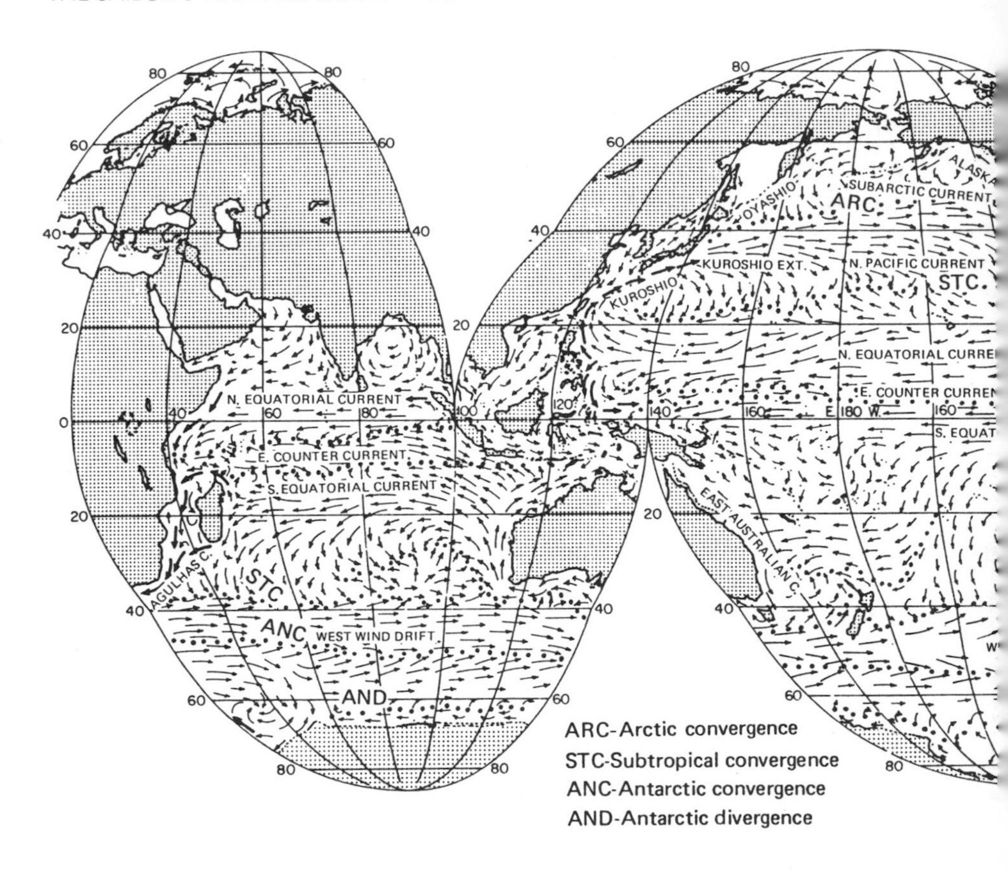

WIND-DRIVEN CURRENTS

Another profound effect of the atmosphere on oceans is the powerful influence of winds on surface currents. The fact that the major surface currents are wind driven can be easily seen. Compare the normal world ocean currents for the month of February in Figure 6-6 with the world wind pattern for the same month in Figure 6-7. For example, clearly the North Equatorial Currents in both the Atlantic and Pacific oceans (Fig. 6-6) are aligned with the Northeast Trades (Fig. 6-7). The Gulf Stream and the North Pacific Current just as clearly follow the prevailing westerlies of their latitude.

The cause of the California Current is also wind even though when you examine Figures 6-6 and 6-7 that current seems to be moving at about right angles to the westerlies. What happens here is that the prevailing winds blowing up against the lee coastline causes an upwelling of cold water that then moves southward along the coast. That cold water then causes frequent West Coast advection fogs as mentioned in Chapter 5.

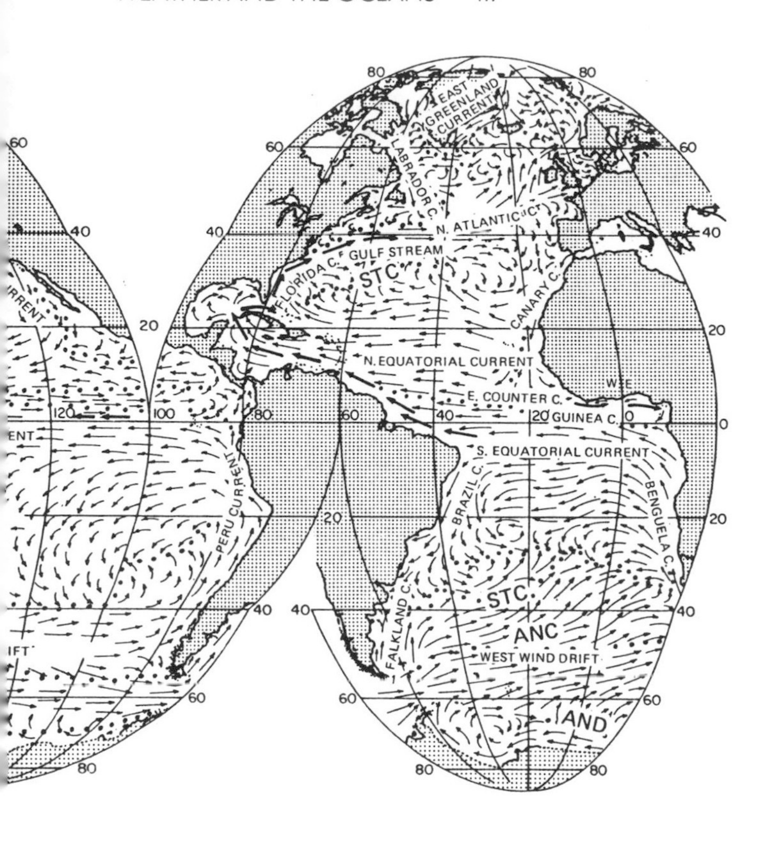

Figure 6-6. LEFT. Surface ocean currents in February. From: Sverdrup, *Oceanography*.

Figure 6-7. BELOW. Ocean surface winds in February. From: Sverdrup, *Oceanography*.

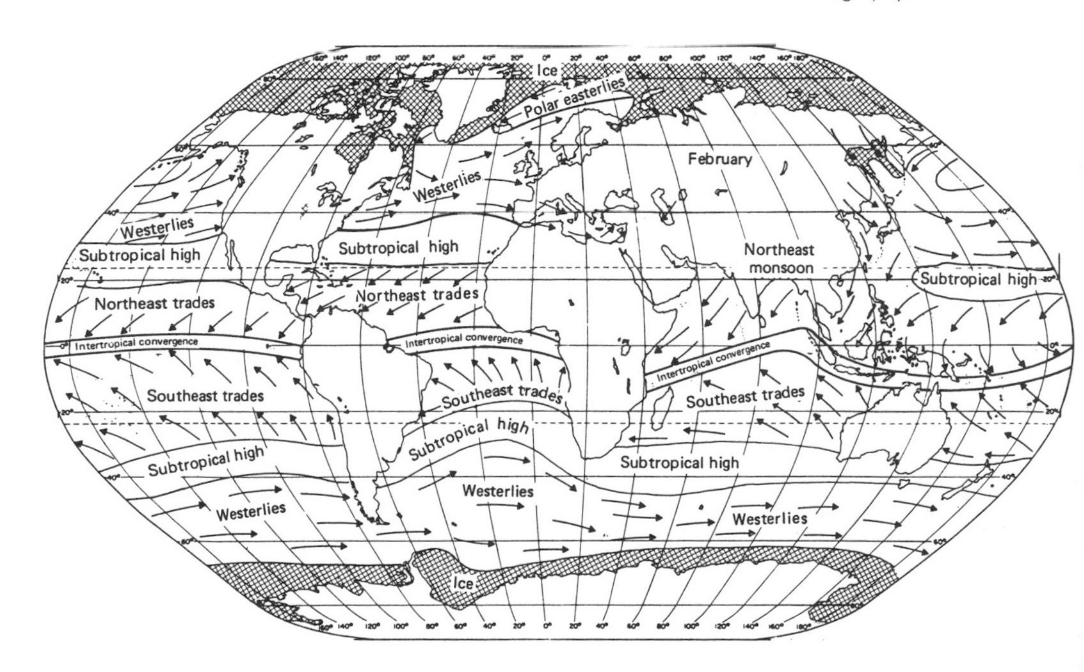

In addition to the large, steady surface currents of the open ocean, the winds also cause numerous minor local surface currents. Some of these become particularly strong and troublesome when reinforced by the daily tidal currents. These same tidal currents may also cancel them out part of the time. Although a few local currents of this type are mentioned in the Coast Pilots, extensive familiarity with local conditions is usually the only way to become fully alerted to them.

BOAT HANDLING IN WIND, WAVES, AND SWELLS

Handling a boat under adverse conditions of wind and sea is probably the greatest test of a skipper's knowledge and skill. The size of the boat is not of primary importance. What is important is matching the boat to conditions for which it is suitable.

A small houseboat of the type in Figure 6-8 is designed for use in rivers, lakes, or sheltered bays, not in open ocean. As long as she is kept in such waters, all will be well. The fine old schooner in Figure 6-9 is an able deep sea boat fully capable of making an ocean passage. She will do well so long as she is not asked to operate in the kind of shoal water that is home to the houseboat. Each has its uses, and its limitations.

FOUL WEATHER

Rain, snow, sleet, and cold—we'll all probably agree that these constitute "foul weather." But how much wind and sea does it take to qualify as foul weather? That varies depending on where you are. On a large, shallow bay or lake, such as much of Chesapeake Bay, or Lake Erie, a fairly moderate wind will build steep, choppy seas and very uncomfortable conditions quite quickly. By contrast the same wind blowing over deeper waters might produce good sized waves, but in the form of slow, rolling trochoidal swells over which even quite small boats will ride easily.

What constitutes foul weather for your boat in your area is probably very clear to you by now. It is also certainly clear to you that taking the boat out in rotten weather conditions is seldom fun. However, the most cautious of us get caught at sea in foul weather from time to time. When this happens there are procedures you can follow that will minimize both personal discomfort as well as any danger to the boat.

Figure 6-8. This houseboat is ideal for use in sheltered bays, lakes, and rivers, but not in open ocean.

Figure 6-9. This classic schooner is well suited for use in open ocean, but not in shallow and constricted waters.

PRELIMINARY STEPS

The prudent skipper keeps a constant eye on the weather. When wind and seas begin to build, or the cloud trains take on an ominous look, or both, prepare for trouble well before it strikes.

- Close and secure all hatches and ports
- Fix and chart your position as accurately as possible
- Stow away all small things and lash down all heavy gear
- Pump out bilges—loose water reduces boat stability
- Break out emergency equipment: life preservers, hand pumps or bailers, sea anchor, flares

HEAD SEAS

A sailboat under sail will be unable to head directly into the seas. The wind will have to come from somewhere in the same quadrant as that from which the seas are coming. Therefore, let us look at how to handle power boats in head seas.

When running head into old, round-topped trochoidal seas (see Fig. 6-4), control of the boat is seldom a problem even if they are great big ones. It may seem like a roller-coaster ride, but it will be a fairly smooth one. It's the steep, pointed cycloidal waves that cause trouble.

Running into steep cycloidal seas, the first thing to do is slow down. Let the boat rise to the seas, not drive into them. Even then you may have to fall off so as to take the seas at an angle somewhere between 30° and 45°. The combination of the rolling and pitching motion this produces is extremely unpleasant as well as terribly tiring, but safer. It could also make some of your passengers or crew deathly seasick. However, it has two major advantages. The hull takes a terrible pounding when heading straight into steep seas. Taking them at an angle eliminates most of this strain on the hull. Also, heading directly into steep seas can lift the propellor completely out of the water allowing it to race. Racing will damage both the engine and the entire drivetrain if allowed to continue for any length of time. Heading off will eliminate this difficulty as well. Reaching your objective against steep head seas in a power boat may thus require "tacking" against the seas in much the manner of a sailboat.

Tacking upwind against steep seas in a sailboat is going to be quite uncomfortable also, but there will be one huge advantage over the power boat: the wind pressure on the sails all but completely eliminates the rolling component making the motion of the boat almost entirely a pitching movement, which is far easier on the crew than a power boat's rolling motion.

FOLLOWING SEAS

The two big dangers when running before a following sea are *broaching,* and *pitchpoling*. Broaching (Fig. 6-10) occurs when the boat slews around, or yaws, so as to lie broadside to the waves which could then roll her over. Pitchpoling (Fig. 6-11) happens when a boat runs down the forward slope of a wave too fast, and drives her bow into the trough. The following wave may then flip her end over end.

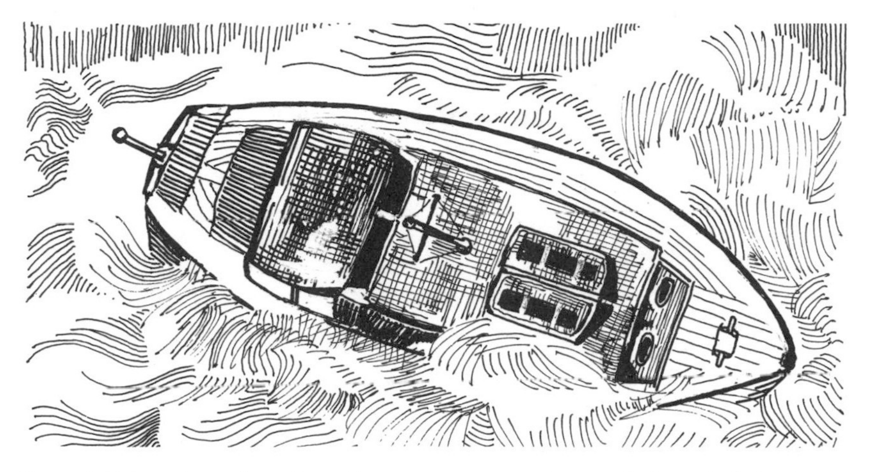

Figure 6-10. Boat broaching. Vessel has swung so as to be caught broadside by the seas.

Figure 6-11. Boat pitchpoling. Wave has lifted the stern, driving the bow into the trough.

Again slowing down is the answer. To achieve sufficient slowing it may be necessary to throw out a sea anchor or drogue astern (Fig. 6-12). This is necessary if slowing the engine leaves the boat with a propellor discharge stream passing the rudder that is too weak to provide steerage. Pulling against the drogue will then restore control.

Hull shape becomes a factor when running before the seas. A double-ender neatly cuts a following sea. Yawing is no problem. The typical broad, flat power cruiser transom is an invitation to yawing and subsequent broaching. In a hull with a wide, flat transom, taking following seas on the quarter rather than dead astern is generally safer. Instead of attempting to run straight before the seas try "tacking" before them.

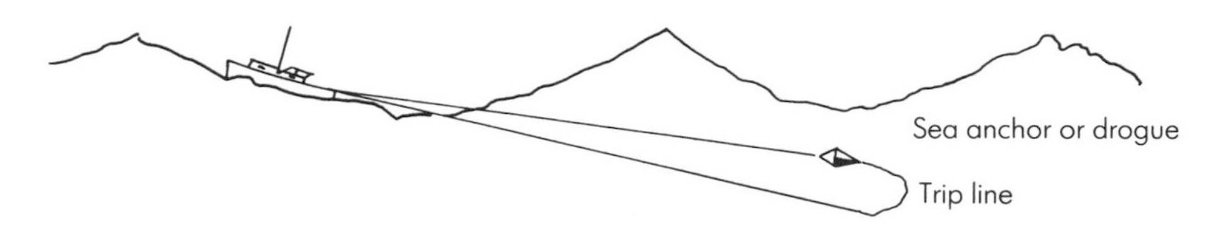

Figure 6-12. Using a sea anchor or drogue to slow progress in a following sea.

ENTERING AN INLET

As the long waves of deep water reach an inlet, the decreasing water depth shortens and steepens them. Also refraction tends to straighten them across the inlet opening as well. No matter how calm it seems outside, always be ready for some surging in an inlet.

When there are fair-sized swells running outside, depend on very turbulent conditions in the inlet. In some cases waves will be breaking all the way across, while in others only part of the way. Under such conditions you should consider entering only if you must—if for good and sufficient reasons you cannot wait outside until the seas moderate.

Having decided to enter, the first thing to do is stand off and watch the wave action for a time. The waves will pass in groups that average about the same height punctuated by one bigger than the rest immediately followed by one smaller. A group may consist of four, or five, or six, or whatever number. It will not be less than three.

After determining the cadence, wait for one of the big ones, and be ready to follow it in. The next few minutes will be critical. They will call for very skillful handling of both the wheel and the throttle. After the big wave has gone in, get halfway up the back of the small one that follows it, and *stay there all the way into the channel.* If the boat goes faster than the wave and tops its crest it is in danger of broaching or pitchpoling. If it goes too slowly and drops in the trough it will be pooped by the following wave. Either one can mean serious trouble.

HEAVING TO IN OPEN WATER

A harbor entrance may appear dangerous making it necessary to stay outside until the seas moderate. Or conditions may become so bad while en route that the boat can no longer make headway without taking severe punishment. In either case it is time to heave to.

A power boat can heave to for a limited time with engines running just enough to provide steerage way to keep her head into the seas. However, as long as this goes on she is burning fuel.

If the boat will have to heave to for an indeterminate length of time, set a sea anchor or drogue from the bow. Then shut off engines as well as all unnecessary electrical equipment so as to conserve both fuel and battery power.

A sailboat can heave to under shortened sail, or she can take all sail down and ride to a sea anchor as well. In any case, when heaving to, be sure there is plenty of "sea room." This means be well offshore. Any boat while heaved to will drift. It will make some sternway as well as some leeway.

Even when there seems to be lots of room keep a good lookout when heaved to. The drift may be faster than anticipated, or in a different direction than expected (unknown current acting). Also, there may be other boats moving around out there either under way or heaved to. Keeping a sharp lookout will enable you to take evasive action should this become necessary.

7

INLAND FRESHWATER LAKES AND RIVERS

As we know, the oceans, the Great Lakes, and other large bodies of water such as the Mediterranean Sea or the Caspian Sea, profoundly influence the climate as well as the day-to-day weather of adjacent land areas. Smaller lakes and rivers also have their effects in a more limited way on local weather. Their influence varies considerably depending partly on the size of the body of water, and partly on the geography of the surrounding land.

Ocean sailors often have a tendency to look down their noses at those who operate boats on the lakes and rivers. Many old "salts" seem to feel that freshwater operation is just too tame for "real" sailors. They either forget, or more likely are completely unaware, that lakes and rivers present some of the same problems they face at sea, as well as some additional difficulties totally unknown to the saltwater sailor. Try putting an old salt encrusted "shellback" on a rubber raft, and then run him through some rapids on even a class 3 or class 4 river. You can depend on getting his undivided attention very quickly! At any rate, for *this* old salt it was certainly an eye opener and a very exciting experience.

As was discussed in Chapter 6, a fairly moderate wind blowing over shallow bays and harbors kicks up very nasty, choppy waves. Exactly the same thing happens on shallow inland lakes. It won't take long to discover that bouncing around on choppy fresh water is just as uncomfortable as bouncing around on choppy saltwater.

The saltwater sailor will also rapidly learn that he cannot see through fog any better when he's on a lake than he could when he was on Chesapeake Bay, San Francisco Bay, or the North Atlantic Ocean. And to make things a little worse, he has no charted aids to navigation with foghorns, gongs, or bells to help him find his way when visibility closes down (except on the Great Lakes).

River currents may also present problems not faced by the saltwater sailor. Except in tidal estuaries, they flow one way only, they do not reverse, and there is no such thing as slack water. Velocities can become tremendous. They sometimes flow at speeds in excess of 10 knots. They are not affected by the winds as ocean currents are, but they are often greatly affected by rainstorms that may occur far out of sight upstream. This may cause sudden and hazardous changes in water volume, water depth, and current velocity totally without warning.

WEATHER INFORMATION

Weather Service VHF-FM marine broadcasts covering inland freshwater lakes and rivers are virtually nonexistent except for the Great Lakes. There extensive coverage is available as shown in Figure 2-3. Additional coverage of the Hudson River, New York State Canals, and Lake Champlain, all of which connect to the Great Lakes, is provided by:

KIG–60	Burlington, VT	162.40 MHz
WXL–31	Syracuse, NY	162.55 MHz
WXM–31	Elmira, NY	162.55 MHz
WXL–34	Albany, NY	162.55 MHz

On most inland lakes and rivers, the boatman is dependent for weather information on commercial radio and TV, and local newspapers. These media will keep you well informed as to the movements of major weather systems in the area, but the continuous updating of local conditions on the water that is available to the coastal navigator on VHF-FM is just not there for the freshwater boatman.

LAKE WATER LEVELS

To the boatman operating on lakes, weather has a huge effect on water depths. This effect is largely seasonal. The lakes are fullest after the spring runoffs which often flood areas that are dry by mid-summer. Since there are no charts, only local knowledge and experience can guide the lake sailor as to where shoals will develop as the lake level changes.

Trimmer Lake on the Kings River in California is a typical example of a good-sized lake extensively used for recreational boating. The photograph taken in mid-July (Fig. 7-1) shows the change in water level between the spring high and mid-summer. It has already dropped about 40 feet, and will continue to drop until the start of the autumn rains.

During a particularly rainy summer, lake levels will stay higher than normal, while during a dry summer they will fall more quickly than usual. However, a single large summer rainstorm is unlikely to cause a sudden, dramatic change in the water level of a lake. By contrast, in a river, a storm upstream can raise the water level several feet in as many minutes!

LAKE WINDS

Winds over lakes, particularly smaller lakes in hilly country, are usually highly erratic and variable. Both velocity and direction can change dramatically over a distance of no more than a few yards. The contours of the land in one place may lift the wind high above the water, or block it, creating a flat, calm spot on the lake. Nearby a valley in the land may funnel the breeze producing a gusty spot alongside the calm one. Handling a small sailboat during a breezy day on a lake in hill country is a lively exercise indeed!

THUNDERSTORMS AND SQUALLS

The thunderstorms and line squalls that accompany the passage of weather fronts can be quite dangerous to lake boaters because of the strong, gusty winds associated with them. All too often, because of intervening woods and hills, oncoming disturbances of this type cannot be seen by the freshwater boater until they are nearly upon him. When newspaper, radio, or TV weather predictions indicate that thunderstorms or a frontal passage are expected, keep a particularly sharp lookout on the part of the sky you *can* see. Also make it a standard operating procedure to pay particular attention to the cloud sequences, and any changes occurring in the direction of movement or velocity of clouds or winds.

Very few boats actually get struck by lightning while out on lakes; however, if you should find yourself out in the middle of a good-sized lake in a thunderstorm, it would be comforting to know that you have recently checked the grounding of the radio antenna, or mast—just in case.

Figure 7-1a, b. ABOVE. Trimmer Lake level in mid-July. Trees on the far shore mark the spring high water level. BELOW The lake will drop considerably further before the fall rains begin in September and October.

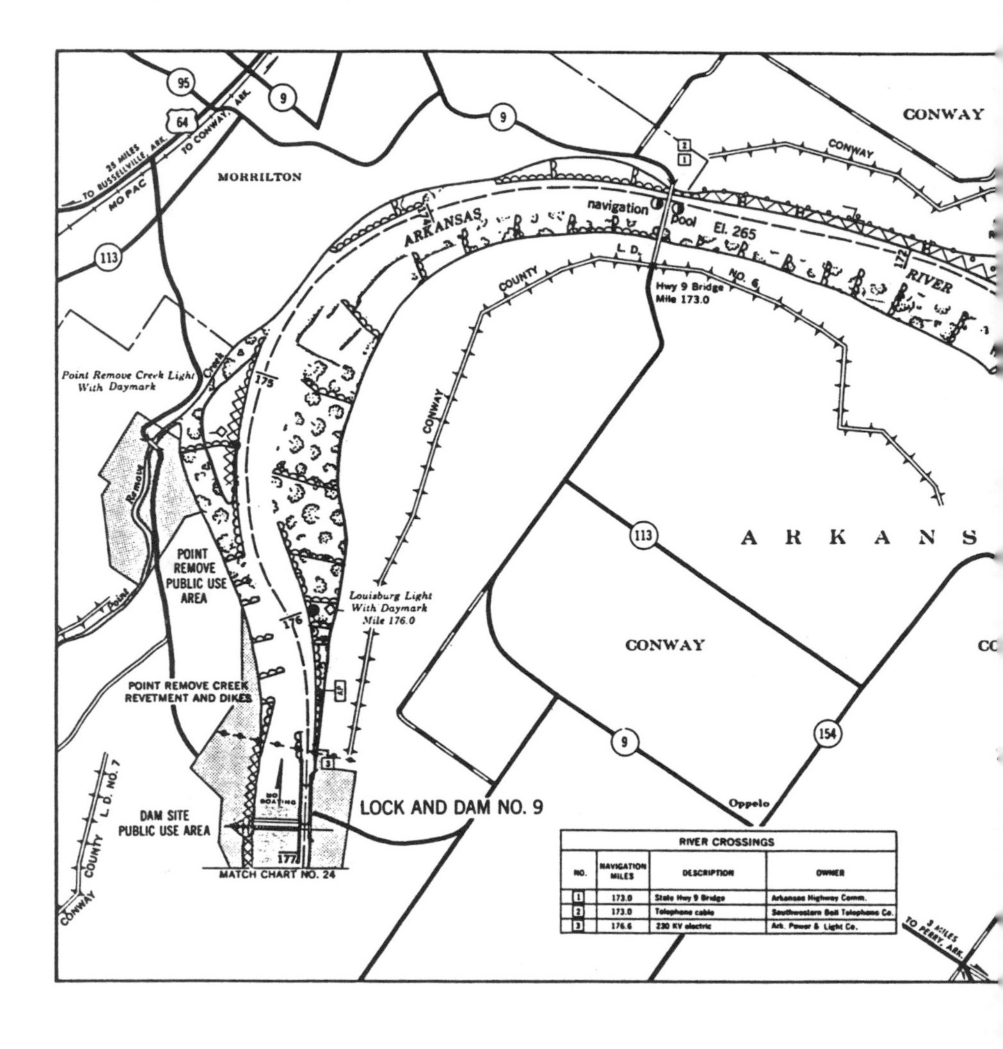

RIVER CROSSINGS			
NO.	NAVIGATION MILES	DESCRIPTION	OWNER
1	173.0	State Hwy 9 Bridge	Arkansas Highway Comm.
2	173.0	Telephone cable	Southwestern Bell Telephone Co.
3	176.6	230 KV electric	Ark. Power & Light Co.

RIVER DEPTHS

Charts do exist for a number of the major river systems that are used for both commercial shipping and recreational boating. One example of this is the Arkansas River Navigation System (Fig. 7-2) which is charted by the Army Corps of Engineers from a point just below Tulsa all the way down the Arkansas River to its junction with the Mississippi. The Mississippi itself and other tributary rivers are also charted by the Corps of Engineers. However, while these charts contain a host of information, very notably absent are the water depths so prominent on the navigational charts of saltwater areas.

Water depths are omitted on river charts because of the magnitude of the seasonal variation. On the Mississippi River at St. Louis the water level changes between late winter and late summer by 40 to 50

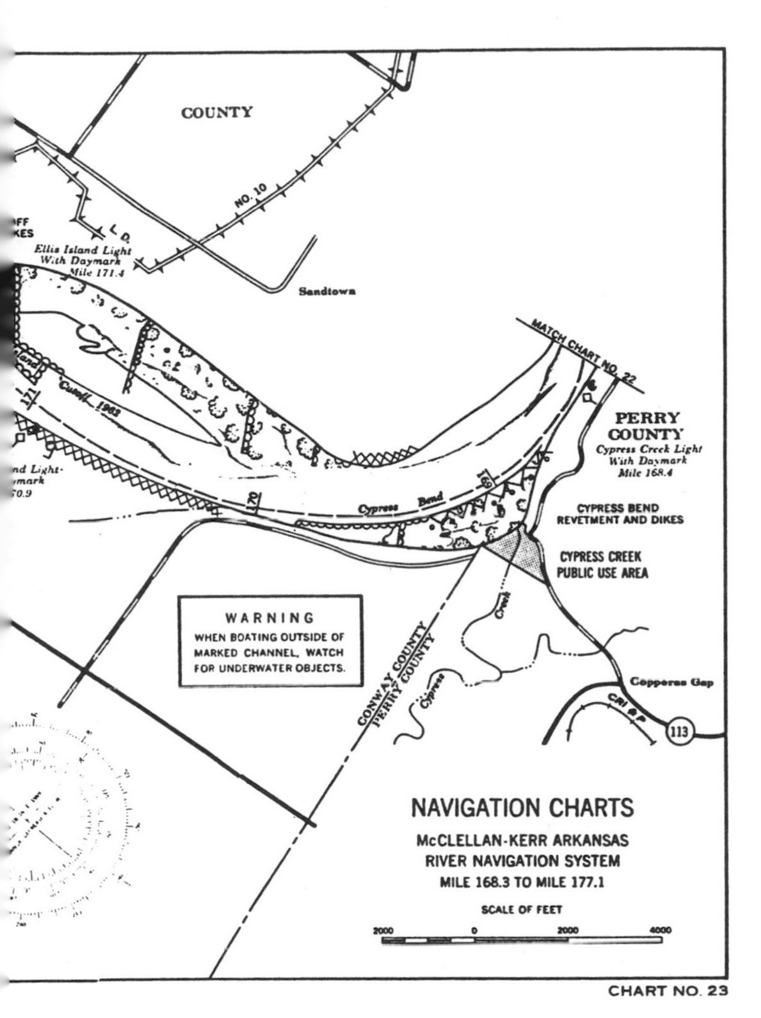

Figure 7-2. Corps of Engineers chart of the Arkansas River Navigation System.

feet. On the charted tributaries, variations of 10 to 30 feet during the same period are common. In addition, sudden summer storms can raise the levels of smaller navigable rivers by several feet in a matter of hours. Peaceful looking mountain streams used for rafting and fishing can become raging torrents in minutes! After the storm water has flowed through, the stream will return to its former peaceful condition almost as quickly.

Although water depths are omitted from river charts, the general information pages included in each book of such charts will note *projected depths* and *channel widths*. These numbers mean that when the river is at its normal late summer to early fall level, there are the channel depths and widths to be expected. The actual level at any specific time must be found by checking the numerous gauges mounted at various points along the river. Gauge locations are noted on the river charts.

RIVER CURRENTS

Unlike ocean currents, river currents are virtually unaffected by winds. In a general way, they vary with the river level, increasing in velocity at high levels and decreasing at lower levels. But the complete picture is actually a great deal more complicated than that. River currents vary in velocity at different places across the width of the river due to irregularities in the contours of the river bottom. The current is normally faster toward the center of a channel, and slower along the sides. It also varies with the width of the channel. It speeds up as a channel squeezes through a narrow gorge and slows down as it spreads out across a wide flood plain. At a bend in a river, the current will usually flow faster around the outer side of the curve and more slowly on the inside.

The strength of a river current varies with seasonal weather changes, but does not significantly change from day to day, with one exception. The same summer storms, mentioned earlier, that can suddenly raise a river level by several feet, will greatly increase current velocity at the same time. After the storm water has passed, the current velocity, as well as the water level, quickly returns to its prestorm state.

Rivers emptying into the ocean are affected for varying distances upstream by the ocean tides and thus by tidal currents as well. Tidal currents in some places reverse river currents for a portion of the tidal cycle. In others, they may temporarily slow or stop the river current. In either case, except as they may affect tides, day-to-day weather changes have a minimal effect on river currents in tidal estuaries.

FOG

Localized radiation or advection fogs periodically form over inland lakes or rivers yet not over the adjoining land. More commonly, fogs that blanket the smaller inland lakes or rivers are included in weather systems covering much larger areas and thus blanket the surrounding land as well. Large areas of fresh water, particularly the Great Lakes, significantly affect and alter the weather systems that pass over them, producing fogs as well as other disturbances (see Chapter 4).

PART III

MECHANICS OF WEATHER

8

HEAT, PRESSURE, AND PLANETARY AIR CIRCULATION

The weather changes that continually move through our atmosphere are powered solely by heat energy received from the sun. The atmosphere absorbs that energy in huge amounts, and with it produces many awesome effects. In simple terms, the atmosphere is an immense heat engine.

THE ATMOSPHERE

A dozen or more gases are present in the air we breathe, but only two account for most of it. By volume, just over 78 percent is nitrogen, and just under 21 percent is oxygen. Argon is a little under 1 percent, and carbon dioxide plus trace amounts of neon, helium, methane, krypton, nitrous oxide, hydrogen, ozone, xenon, radon, and nitric oxide make up the small remaining percentage of fixed componants. The relative proportions of these gases to each other in the atmosphere remains constant out to a height of about 40 miles, which is well beyond the area with which we are concerned.

In addition to these constants, our air contains varying amounts of dust particles, salt crystals, various gaseous and manmade solid pollutants, as well as water vapor. The air content of these variables changes considerably with time and location. The most important influence on weather among these variables is water vapor. This is due to the tremendous heat exchanges involved in the vaporization, and condensation of water.

While immense amounts of water in vapor form are carried in the air, about 90 percent of this vapor is found in the lowest few miles of the atmosphere. Evaporation from the world's oceans is the source of this atmospheric water vapor. As we shall soon see, temperatures at higher altitudes are too cold to allow water to remain in vapor form, consequently water vapor is found only at fairly low altitudes. When this vapor condenses out in the form of water droplets, ice crystals, or snow, it still is limited to comparatively low altitudes. The air above 40,000 to 50,000 feet is simply too thin to support ice crystals or snow in suspension.

HEIGHT AND STRUCTURE OF THE ATMOSPHERE

The atmosphere of mixed gases that surrounds our planet decreases in density very rapidly with altitude. It also, with some important transient exceptions, steadily decreases in temperature with altitude for at least the first 30,000 feet to 40,000 feet (Fig. 8-1). At the level of the stratosphere where temperature starts to increase, the air has become highly rarified. The weather phenomena with which we are concerned all occur well below this altitude.

As sailors, we live along the meeting line (or "interface" if you prefer the more pompous terminology) between the water and the air. Sometimes forgotten is a major difference between the two substances. The water on which we sail is almost totally incompressible. Its density cannot be substantially altered by pressure. Air, on the other hand is extremely compressible.

This is the reason for the higher density of air at sea level, and its decrease in density with altitude. Where you reach an altitude of only about three and one half miles above sea level, fully half of the earth's atmosphere is below you! The troposphere averages five miles in height above the poles, increasing to about 11 miles above the equator. It is in the layer of heavy air in the lower *troposphere* that most of the weather changes which affect us, and our vessels, take place.

HEAT ENERGY

The sun generates and radiates electromagnetic energy over a vast range of wavelengths. Its visible light is only a small part of its total radiation spectrum (Fig. 8-2). However, most of the heat produced by solar radiation happens to fall in this same visible band (Fig. 8-3).

When this light strikes the earth, part of it is immediately reflected away as light, another part is absorbed by the material it struck as

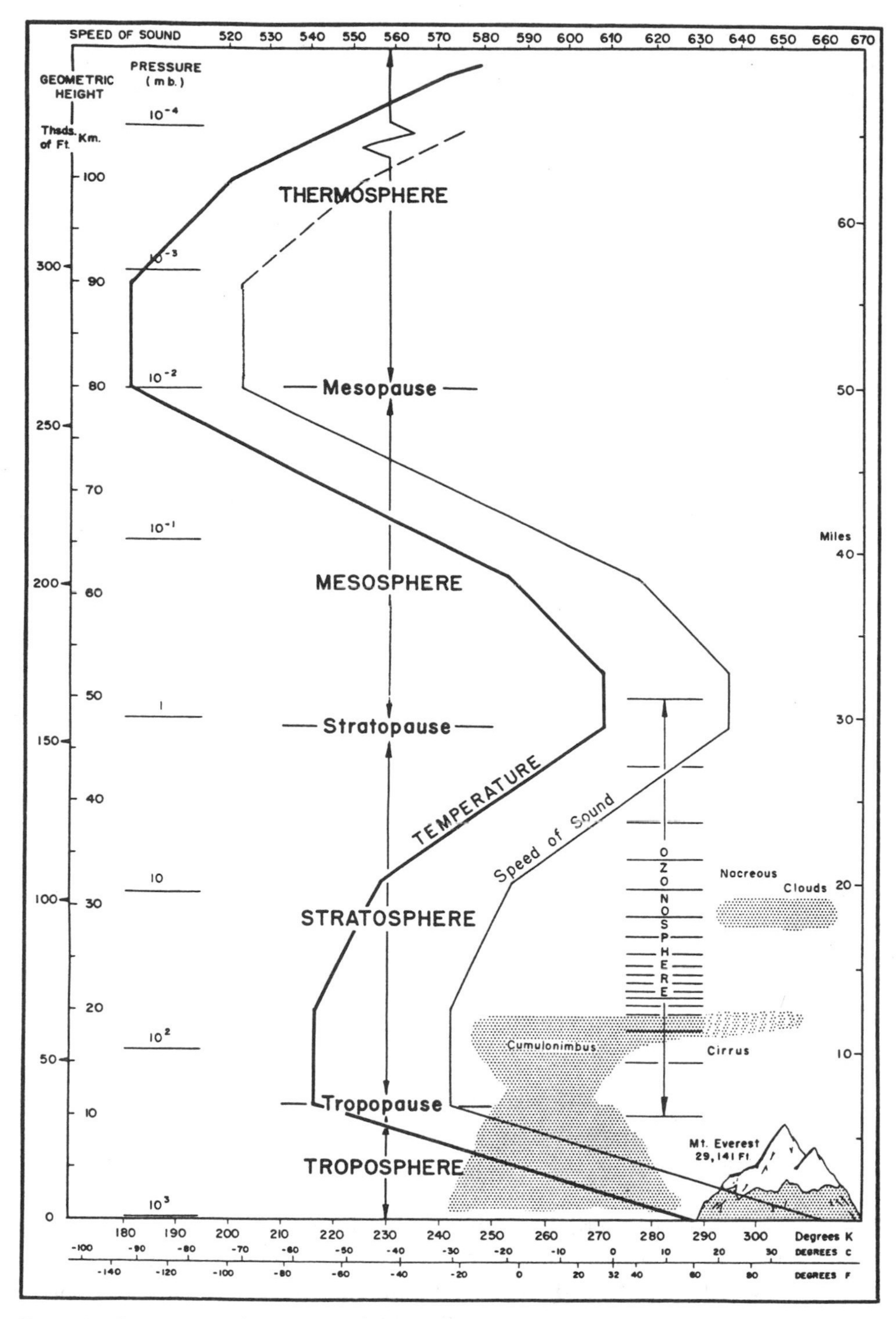

Figure 8-1. Temperature changes with altitude in the atmosphere.

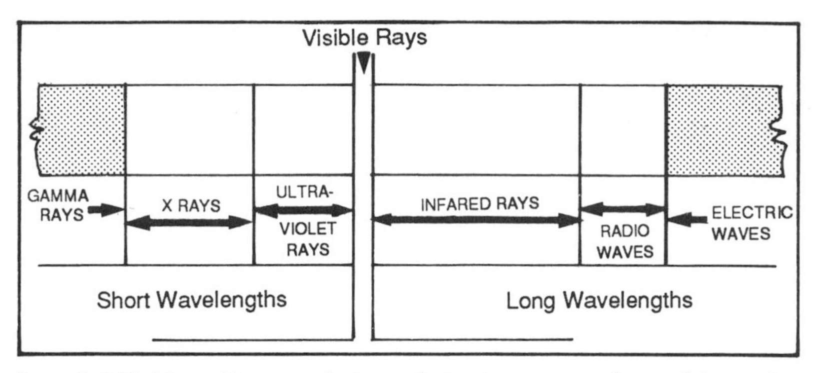

Figure 8-2. Visible and heat-producing radiation is a very small part of the total range of solar radiation.

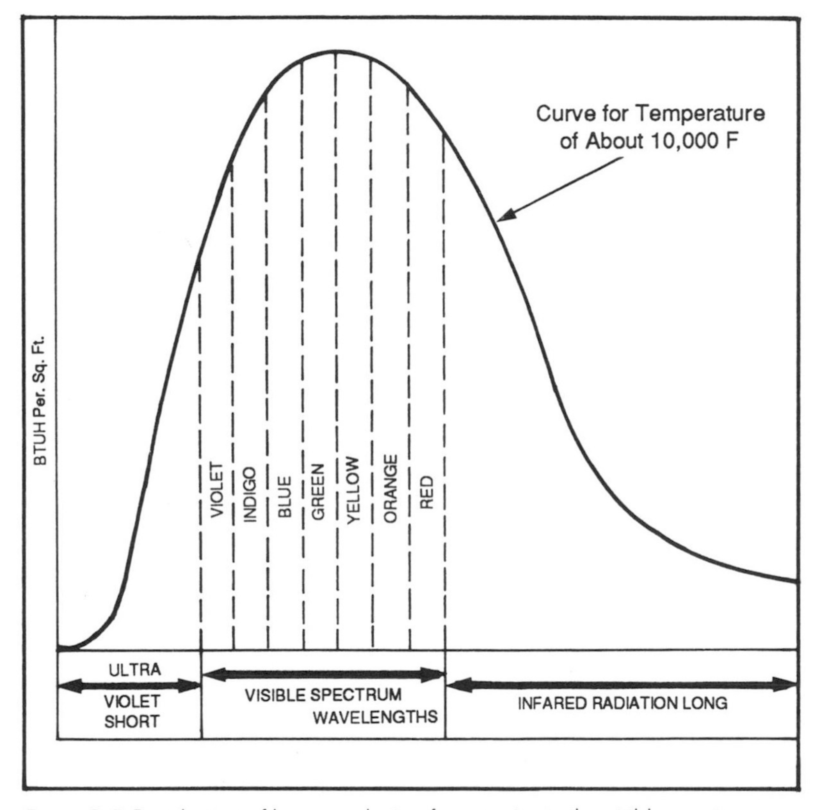

Figure 8-3. Distribution of heat-producing frequencies in the visible spectrum.

heat. A third fraction is transformed into long wave infrared heat radiation. The part reradiated as long wavelength infrared heat is very different from the short wavelength visible solar light. Visible solar light waves pass readily through glass. Infrared waves do not, nor do they pass through clouds readily either (Fig. 8-6). This blocking of infrared heat by cloud decks is termed the *greenhouse effect*.

The greenhouse effect is very significant in warming temperatures below cloud layers, and also in dissipating some types of clouds by heat from below (see Chapter 9).

HEAT AND TEMPERATURE

While *heat* and *temperature* are closely related, their meanings, although quite distinct and different, are often confused. Temperature is comparatively easy to measure, and therefore to define. Anything that is hot has a high temperature which will show up on an appropriate measuring instrument.

Heat is not as easy to define or to measure. If a one pound rock and a one pound pan of water are heated by two identical burners the temperature of the rock will rise much more rapidly than that of the water. Although the temperature, size, and heat output of both burners is exactly the same, due to the different physical characteristics of the two materials it takes more heat to raise the temperature of the water than it does to raise the temperature of the rock.

Water is a particularly great "heat sink." This means that a comparatively large amount of solar radiation is required to noticeably raise the temperature of a body of water. By contrast a fairly small amount of radiation falling on a sandy beach, as doubtless you have often observed, can quickly make that sand uncomfortably hot to your bare feet!

Once warmed, water will hold its temperature far longer than sand, rock, or dirt because it had to absorb far more heat in order to raise its temperature in the first place than did the sand, rock, or dirt. This difference in heat absorption and retention between the land and the water has profound effects on air temperatures, winds, and weather.

HEAT TRANSMISSION

In order to raise the temperature of any substance, heat must be transmitted into it from something else that is hotter. Heat is transferred from one substance to another by any one, or a combination, of three methods: radiation, conduction, or convection.

RADIATION

Heat transferred by radiation is transmitted in the form of electromagnetic waves. No transmitting medium is required. The heat received by the earth from the sun is just such radiant energy that has passed in wave form through space.

Heat that is felt when holding your hand a few inches above a burner on an electric range is largely long wave infrared heat rays radiated by the burner. Only a small part of what you feel is caused by air warmed through contact with the burner.

Similarly, much of the heat felt when standing close to a steam radiator is caused by radiant heat waves, as is much of the heat felt when sitting close to a crackling fireplace in the winter. A small proportion of the heat felt in these instances results from one or both of the other two methods of heat transfer.

CONDUCTION

Conduction is the transfer of heat by means of direct contact. Heat may be conducted through a single material from one part to another, or heat may be conducted from one material to a second one by placing them in direct contact with each other.

If one end of a metal bar is placed in a flame, fairly soon the other end will become hot. In this instance heat is conducted through the material itself from one particle of metal to another.

Air that is in direct contact with earth, sand, or rock that has been heated by the sun is heated by direct conduction. Warm air passing over cold water, ice, or snow loses heat also by the same mechanism of conduction. Conduction works both ways. The substance that is cooler gains heat from the warmer one. In the process the warmer one loses heat. This process also has important consequences on weather.

CONVECTION

In relation to weather phenomena, the most important form of heat transfer is convection. In this process heat is transmitted through the physical movement of heated material. When the air above a radiator in your home is heated, it expands, becomes lighter than the air around it, and rises. This allows cooler air to move in over the radiator where it also is heated, expands, and rises. This process becomes continuous creating a *convection current* that gradually distributes heat throughout the room.

On a vastly larger scale, air warmed by contact with the earth expands and rises. This vertically moving volume of air is replaced by horizontally moving air that we feel as wind. The various weather phenomena we encounter from day to day are caused basically by the distribution, and redistribution, of heat through the atmosphere as the result of vertical and horizontal movements of air. The formation and dissipation of clouds as well as the formation and later disintegration of major storm systems all involve massive convective movements of heat.

UNEVEN ATMOSPHERIC HEATING

The electromagnetic radiation produced by the sun is substantially constant in intensity, and in distribution of frequencies (see Figs. 8-2 and 8-3). However, the heating of the earth's atmosphere by that radiation is extremely uneven and irregular. Several factors contribute to producing this result.

LATITUDE

The amount of solar radiation that will strike a unit area of the earth's surface varies greatly with latitude (Fig. 8-4). As latitude increases, a given parcel of solar radiation spreads over an increasingly large area of the earth's surface. The intensity of radiation per square foot of surface is correspondingly decreased. The area at right angles to the sun will receive the maximum possible concentration of radiation. As the angle changes with increasing latitude, the concentration per unit of surface decreases.

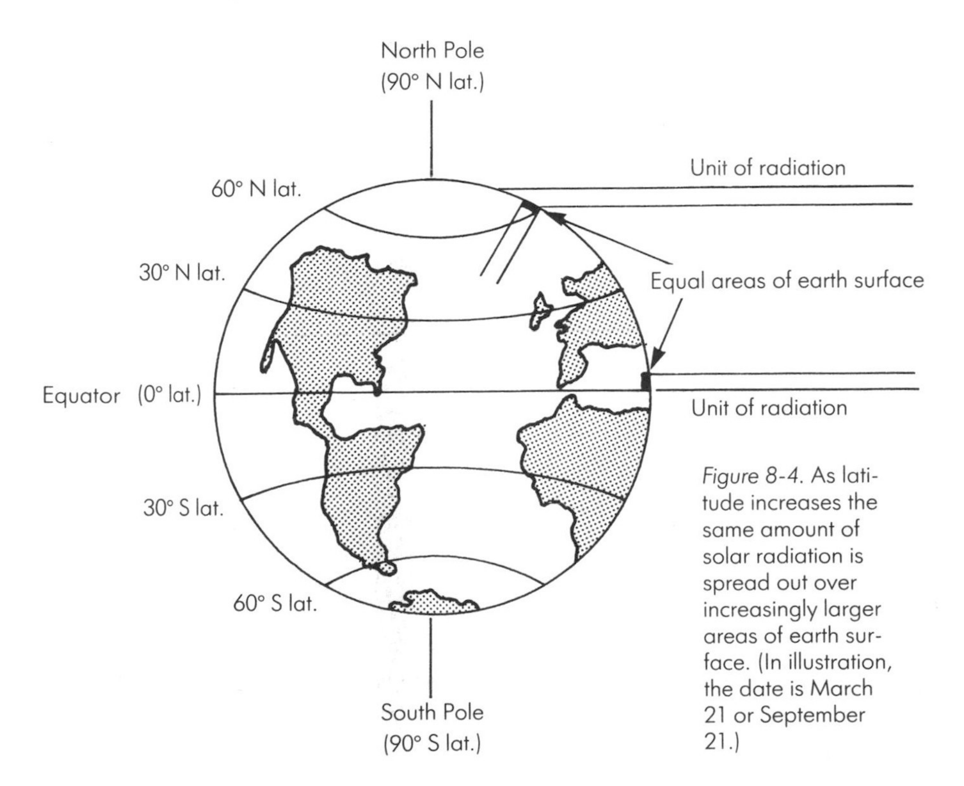

Figure 8-4. As latitude increases the same amount of solar radiation is spread out over increasingly larger areas of earth surface. (In illustration, the date is March 21 or September 21.)

SEASONAL VARIATION

The earth circles the sun once a year following a path called the *plane of the ecliptic* (Fig. 8-5). In the process, it moves around an imaginary center line called the *axis of the ecliptic.* At the same time, the earth continually rotates around its own axis once a day. Just to add some complications none of us need, the axis of the ecliptic and the axis of the earth are not parallel. The two differ by an angle of 23° 27′.

This causes the sun to apparently move north and south with the passing seasons relative to any particular place on the surface of the earth. When the sun is directly over the equator, (see Fig. 3-4) a square foot of earth there will receive vastly more solar radiation than an equal area at say latitude 60 N. However, the sun is only directly over the equator twice a year, on March 21 and on September 21.

By June 21, the sun appears to be directly over latitude 23 N. At that point latitude 23 N gets as much radiation per square foot as the

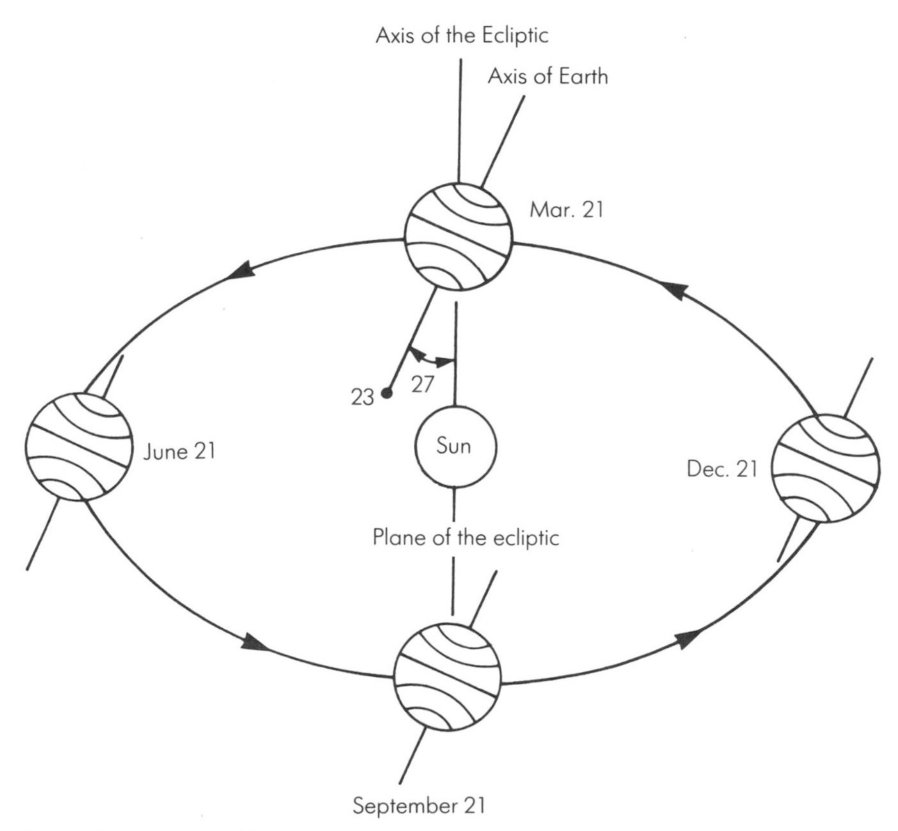

Figure 8-5. Seasonal differences in the angle of the earth relative to the sun changes the amount of heat reaching different parts of earth as the year progresses.

equator did on March 21, and also the radiation per square foot way up at 60 N has increased immensely (by over one third). On December 21, mid-winter in the northern hemisphere, the sun has moved down to latitude 23 S. The equator is getting the same amount of radiation per foot that it got on June 21, but latitude 60 N is getting much, much less!

The atmosphere gets far more of its heat as reradiation from the earth than it does through direct radiation from the sun. Consequently, seasonal changes in the heating of different areas of the earth's surface are paralleled by variations in the heating of the atmosphere above those areas.

TIME OF DAY

At different times of day we again find differences in the angles at which the sun's radiation strikes any particular part of the earth. As the sun crosses the sky each day, its angle relative to any place on the surface changes constantly. This means the intensity of its radiation hitting that place also changes constantly. Here again uneven surface heating results in uneven heating of the atmosphere above it.

SKY CONDITION

By sky condition we mean the amount of cloud cover. This, as you have certainly noticed, varies not only from day to day, but also from hour to hour during a day. An overcast sky significantly reduces the intensity of direct radiation received at the surface during the daylight hours. However, as we shall see in further discussion of the greenhouse effect in Chapter 9, an overcast sky at night significantly reduces heat *loss* to outer space during the hours of darkness. By decreasing heat gain during the day, while also decreasing heat loss at night, an overcast sky has an overall effect of reducing rather than increasing atmospheric temperature variations.

ATMOSPHERIC CONDITIONS

The intensity of radiation reaching the earth is also affected by the clarity, or lack of it, of the atmosphere itself. Is visibility good, or is it reduced by haze, smoke, dust, or other suspended material? No one will be surprised to find that air through which you cannot see clearly also reduces the intensity of the solar radiation passing through it (Fig. 8-6). The same factors that reduce visibility tend to increase the greenhouse effect. In turn, this also has a leveling and stabilizing influence on atmospheric temperature.

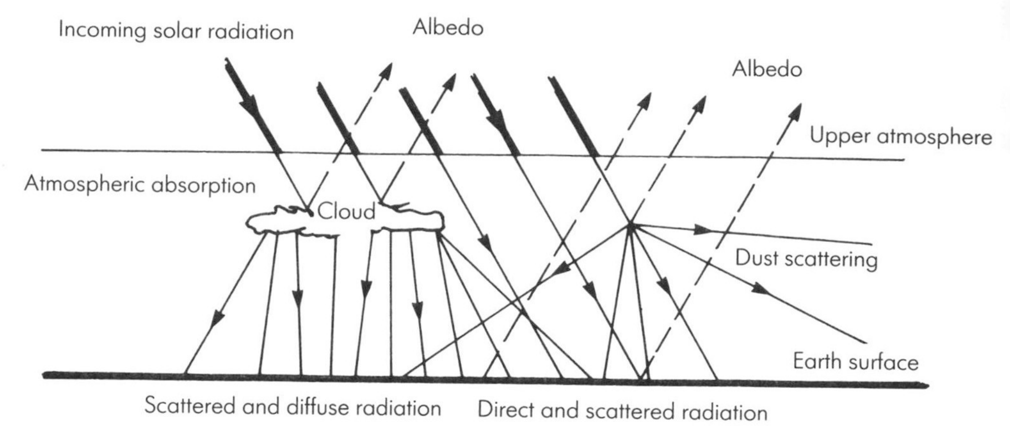

Figure 8-6. Distribution of solar radiation that reaches earth's atmosphere.

ALBEDO

As mentioned earlier, the vast majority of the heat energy that comes to our planet is shortwave electromagnetic radiation lying in the visible part of the spectrum (see Fig. 8-2). The atmosphere absorbs a little of this radiation as it passes through, up to perhaps 17 percent. The earth itself absorbs some more, and a third portion is reflected back out into space. The portion of the radiation reflected back to space is termed *albedo.*

A large proportion of the solar radiation intercepted by the clouds is reflected back into outer space. Dust or other matter in the air also reflects radiation. Of what reaches the earth's surface, part is absorbed, and part reflected back upward. The amount of absorption versus reflection varies with the color and character of the surface. Following are some typical values for albedo expressed as percentages of incoming radiation that will be reflected back to space:

SURFACE	90 REFLECTED
Cloud	50 to 80
Snow—Grassland	70 to 80
Snow—Timberland	40 to 50
Ice	50 to 70
Grassland	15 to 30
Timberland	3 to 10
Water areas	2 to 5

Of the radiation absorbed by the land areas most is used to heat only the upper few inches of the surface because earth is a poor conductor of heat. During the day, the warmed earth surface heats the air in contact with it. At night, that surface rapidly cools, in turn cooling the lowest layer of air.

Over bodies of water—oceans, lakes, bays, and rivers—the situation is quite different. While water absorbs a huge proportion of the radiation it receives, its surface temperature remains substantially constant both day and night. While the actual absorption of radiation by water occurs mostly in a shallow surface layer, that heat is distributed through much deeper levels due to the mixing action of wind, waves, and currents. The end result is very limited variation in the surface temperature between day and night, or between cloudy and clear days. The evenness of the water surface temperature tends to limit the temperature variations of the air in contact with it.

HUMIDITY—LATENT HEAT OF VAPORIZATION AND CONDENSATION

Evaporation from oceans and other large bodies of water, and the subsequent condensation of that vaporized water into cloud droplets, rain, or snow, requires tremendous heat exchanges. To understand the magnitude of these exchanges, and how they affect the weather, we need to look at the mechanisms involved in evaporation and condensation.

Water, the chemical compound H_2O, can exist in any one of three forms, or states. It can be a solid such as ice or snow. It can be a liquid, or it can be a gas in the form of water vapor mixed with the air. To change from the solid state to liquid, or from liquid to vapor requires the absorption of huge amounts of heat.

Heat and temperature, as has been noted, have different meanings. Temperature is found by simply reading an appropriate thermometer or gauge. Heat is more difficult to pin down and measure, but in a roundabout way it can be done. One common unit for measuring heat is the British Thermal Unit commonly referred to as the BTU. One BTU is defined as the amount of heat required to raise the temperature of one pound of water by one degree F.

That definition is likely to be completely meaningless to you. How often do you think of water in terms of pounds? Until now probably never. Then try thinking of a little container 3 by 3 by 3 inches. Fill that and you have about one pound of water. Or perhaps better yet think of filling about three and a half dinner glasses. Enough heat to raise the temperature of that much water by one degree F is one BTU.

We have no interest in the BTU for purposes of making quantitative heat measurements. We simply want a frame of reference for understanding the magnitude of the heat exchanges that occur in the atmosphere as water changes state between solid, liquid, and vapor.

To raise the temperature of one pound of water by one degree F took one BTU. To change that same pound of water from the liquid state to vapor takes *970 BTUs!* In the process the temperature has not increased at all. The heat has been used entirely to accomplish the change from liquid to vapor form. All the heat is now stored in the vapor. For that reason it is called the *latent heat of vaporization*. Back in high school physics you might also have heard the term *latent heat of evaporation*. It means the same thing.

Here is an example of heat being taken up by evaporation with which you are certainly familiar. Let's say you are lying in the sun by a pool on a hot summer day. You go in the pool to cool off. You then come out to dry in the sun. *While you are drying off* you feel cool because the water evaporating from your skin is taking heat with it. As soon as you are dry you again feel the heat of the sun.

The water that leaves the oceans to mix with the air as water vapor thus takes immense amounts of heat with it. That heat is replaced by additional solar radiation absorbed in huge quantities by the oceans. This absorption is aided by the fact that an ocean has very low albedo meaning it does not reflect much of the radiation that strikes it.

When that vaporized water later condenses into liquid water droplets, as when clouds are formed, the heat picked up during evaporation must be released into the air. Thus, this same heat may also be referred to as *latent heat of condensation*.

The change in state of water between the solid and the liquid forms also involves a large heat exchange, but not nearly as large as between liquid and vapor. To go from ice to liquid water takes 144 BTUs per pound. This is a substantial decrease from the 970 BTUs needed to go from liquid to vapor, but still a significant increase from the one BTU per pound per degree of temperature F.

High altitude clouds are made up of ice crystals. When these descend and melt at lower altitudes, as they sometimes do, forming water droplets or rain, considerable heat is withdrawn from the atmosphere in the process.

The amount of water (or humidity) contained in the air, and the state, or form, of that water is a major influence on air temperatures. This results from the huge quantities of latent heat involved in changes of state occurring between water vapor, liquid water, and frozen water while suspended in the atmosphere.

VERTICAL TEMPERATURE CHANGES

As shown earlier in Figure 8-1, with increasing altitude there is a steady decrease in temperature as one ascends through the lower atmosphere. The rate of this decrease varies from time to time, and from place to place due to the uneven heating of the atmosphere as already discussed. The rate of decrease at a specific place and time is termed the *lapse rate*. With proper equipment, measurements can be taken to determine the lapse rate at a particular place and time.

While the lapse rate varies widely, what is called the *normal lapse rate* averages out at 3.5 degrees F per 1,000 feet. Note that the lapse rate is a profile of the temperature changes in a stationary vertical column of air. This means that no vertical air movement, either up or down, is going on.

As we shall see, particularly in connection with air masses and storm systems, periodically a warm air layer moves over colder air. In this case, the temperature will fall with altitude through the cold layer, but then rise suddenly when the higher warm layer is reached. Within the warm layer, it will then start to fall again. This sudden rise in air temperature with altitude is called an *inversion*. Inversions may also be caused by the cooling of air close to the ground due to loss of heat to a cold earth or water surface.

The normal inversion shown in Figure 8-1 that occurs high in the atmosphere above the tropopause does not concern us. The weather phenomena with which we are dealing all take place at much lower altitudes.

CAUSES OF VERTICAL AIR MOVEMENT

For several reasons volumes of air may be set into vertical movement. When an area of the earth's surface is heated, the air above it is heated by that heated earth. This air expands, becomes lighter than the air around it, and starts to move vertically upward (Fig. 8-7). Correspondingly, air above a cold surface will cool and condense, becoming heavier than the air around it, and tend to sink.

Another cause for vertical air movement is produced when horizontally moving volumes of cool and warm air meet. Cool air is always denser and heavier than warm air. The cool stays under; the warm moves up and over the cool (Fig. 8-8). As we shall see, this is an important aspect in the development of cyclonic storm systems.

Horizontal movements of volumes of air may also result in convergence or divergence (Fig. 8-9). Convergence results in vertically

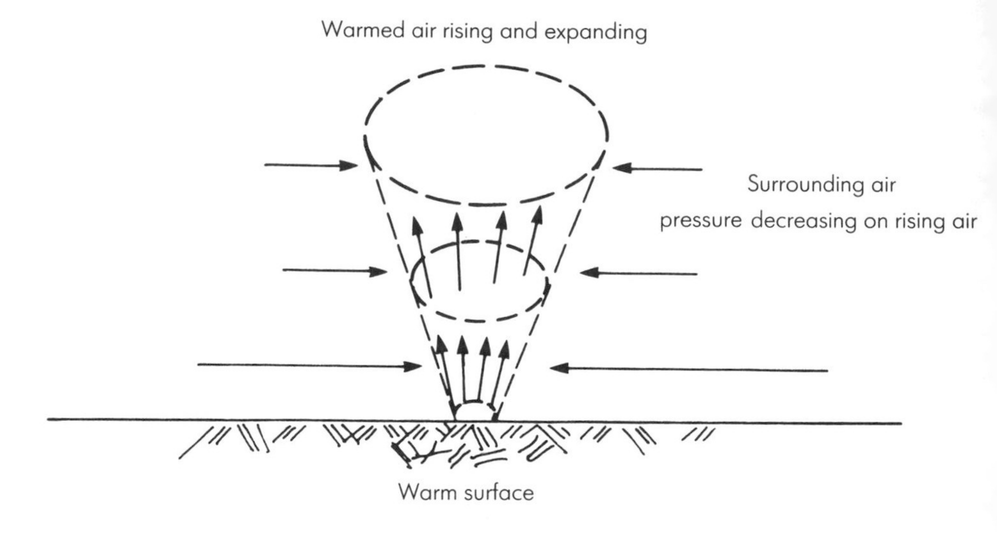

Figure 8-7. Warmed air column rising and expanding through stationary surrounding air.

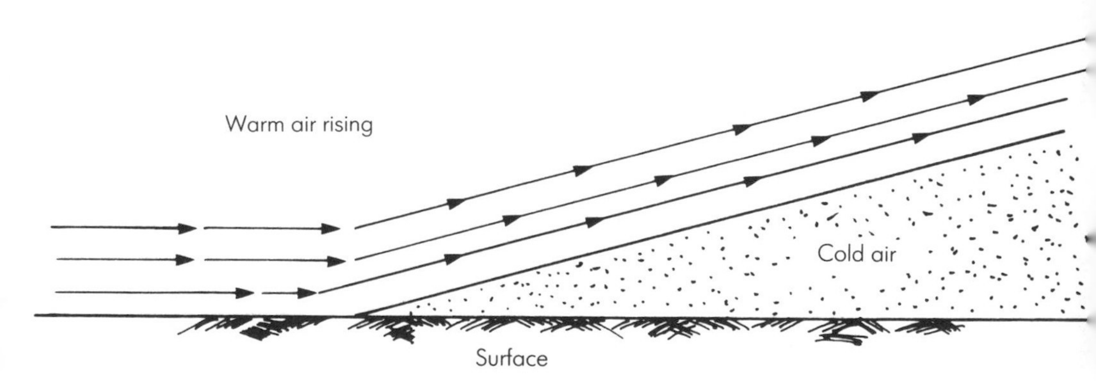

Figure 8-8. Lighter warm air rising over volume of denser, heavier cold air.

rising air currents. Divergence produces vertically descending currents. Later, when we look at planetary air circulation patterns, we shall see the profound effects of convergence at the Intertropical Convergence Zone, and the results of divergence in the "horse latitudes."

Air may also be moved vertically by encountering sloping ground such as hills or mountains (Fig. 8-10). Air forced up in this manner also cools as it rises, and warms again when descending.

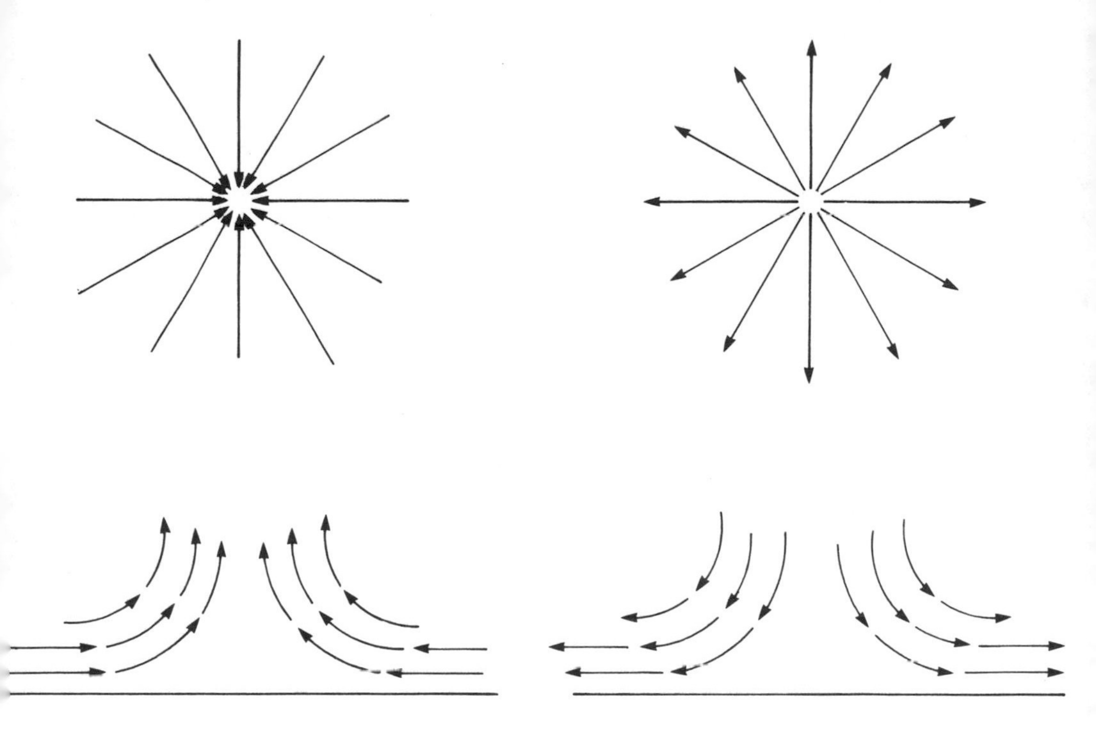

Figure 8-9. Air converging and rising. Air diverging and sinking.

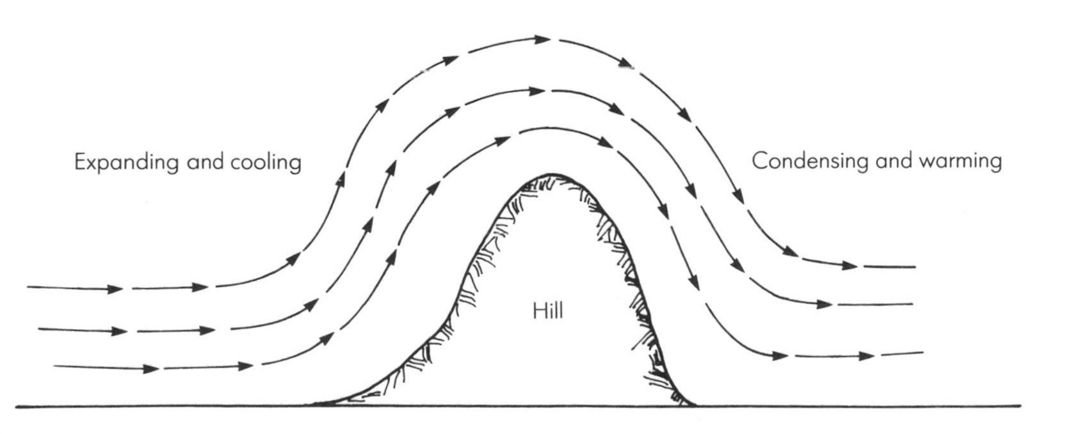

Figure 8-10. Air lifted by passing over high ground.

ADIABATIC MOVEMENT

A column of air moving vertically upward through a stationary volume expands and cools as it goes due to the decrease of pressure with altitude. When descending, a similar column is compressed and warmed. This occurs without the moving air column mixing with the surrounding volume. This form of cooling, or warming is called *adiabatic,* meaning that it results solely from the pressure change that occurs with changes in altitude.

The rate of cooling or warming that occurs under these conditions is called the *adiabatic lapse rate.* As long as the air being cooled or warmed is at less than saturation humidity, its lapse rate will be the *dry adiabatic rate* which is 5.5 degrees F per 1,000 feet. Suppose, for example, some air close to the ground is heated to a temperature of 85 degrees F. It expands and starts to rise. As it rises it cools at the dry adiabatic rate. Assume it reaches an altitude of 2,000 feet. Its temperature by now will have dropped to 74 F ($5.5 \times 2 = 11$ degrees ~ $85 - 11 = 74$)

As air temperature decreases, so does the capacity of that air to hold water as vapor without becoming saturated. When a volume of air reaches saturation temperature, its water vapor starts to condense as liquid water droplets. In this condensation process, the latent heat absorbed during vaporization is released. When condensation starts in adiabatically rising and cooling air, the temperature lapse rate abruptly changes from the dry adiabatic rate of 5.5°F per 1,000 feet to about 3.2°F per 1,000 feet. This is because all of that latent heat being released by condensation (970 BTUs per pound of water) warms the air approximately another 2.3°F per 1,000 feet.

The essentially constant adiabatic lapse rates occurring in vertically moving air are not to be confused with the more variable lapse rates that occur in air that has no vertical movement. The word adiabatic is the key since it applies only to air in vertical motion.

HORIZONTAL TEMPERATURE CHANGES

Due to the uneven atmospheric heating mentioned earlier, numerous horizontal temperature variations are found in the atmosphere. The major overall horizontal change is the gradual decrease in air temperatures moving from the equator toward the poles. This results from the decreasing intensity of solar radiation per unit of surface area (see Fig. 8-4).

While the decrease in intensity of solar radiation is uniform at any specific latitude, its effect on temperature is certainly not. As mentioned earlier, the atmosphere gets most of its heat from the earth. Therefore, land areas that have high surface temperatures, because the heat stays at the surface, raise air temperatures far more than horizontally nearby water areas where the equivalent heat is distributed more deeply. In winter, the same land areas being colder than the sea now keep air temperatures lower than those over the water.

Warm and cold ocean currents also have their effects on altering air temperatures in the horizontal plane. Winter temperatures in London are mild indeed compared with Calgary, Canada, at the same latitude, thanks to the warming influence of the Gulf Stream.

VERTICAL ATMOSPHERIC EQUILIBRIUM

The vertical equilibrium of a volume of air is a measure of its tendency toward vertical movement. Air that is not in movement vertically is considered to be in vertical equilibrium. Balance is a simpler word for it. As we have seen air may be set in motion vertically in several ways. It may be heated or cooled by the earth below it. It may encounter warmer, or colder air through horizontal movement. It may be forced up or down due to horizontal convergence or divergence. It may be forced up by high ground. The extent to which vertical air movement will continue once it has started depends on the sort of equilibrium prevailing in that air.

Air may be in what is called *stable equilibrium,* or it may be in *unstable equilibrium.* To clarify the terms, think for a moment of two cylindrical flashlights resting on a table. One is lying flat on the table, the other is standing on one end. Pick up an end of the one that is lying flat, and then let go. It will simply drop back to where it started. Now tip the one standing on its end somewhat, and let go. It will keep on moving until it is lying on its side. While standing on its end it was at rest. It was in equilibrium, but *unstable equilibrium,* because when set in motion it kept on moving, ending in a totally new position.

An understanding of vertical equilibrium in air brings us back to lapse rates. Specifically, the relationship between the adiabatic lapse rates, and the lapse rate of air that is not in vertical motion determine the stability or instability of a volume of air.

STABILITY

Let us assume that the air in a general area is at 85°F, and the lapse rate in that air is at the "normal" average of 3.5°F. Somehow a parcel of that 85°F air is forced up 1,000 feet. Its temperature has then dropped by the adiabatic lapse rate of 5.5°F to 79.5°F. The surrounding air, meanwhile has dropped only 3.5°F to 81.5°F (Fig. 8-11) which leaves it warmer than the parcel that was lifted. The lifted parcel, being cooler, is heavier, and thus wants to drop back to its starting level. There temperature, and hence density, will be equal. The relationship between lapse rates in this case has produced stability.

If the uplifted air should be forced further on up to 2,000 feet, its temperature will go down by another 5.5°F to 74°F while the air around it will only go down to 78°F. The temperature gap will have widened, increasing the inclination of the uplifted air to settle back.

INSTABILITY

The lapse rate for vertically stationary air, as has been noted, is variable. Take the same example of some air at 85°F, a parcel of which is forced upward. This time the lapse rate of the surrounding air is 6.5°F. At an altitude of 1,000 feet the uplifted parcel has cooled by 5.5°F (adiabatic lapse rate) to 79.5°F again, but the surrounding air cooled by 6.5°F to 78.5. The raised parcel is now warmer and therefore lighter than its surroundings so it continues to rise. At 2,000 feet, the raised parcel is at 74°F, but the surrounding volume is now only 72°F. The gap has widened. The uplifted parcel continues to rise. This is a totally unstable condition (Fig. 8-12).

CONDITIONAL INSTABILITY

In the stable situation shown in Figure 8-11, the uplifted air becomes increasingly colder relative to the surrounding air with increasing altitude. This causes an increasing tendency to sink. In the unstable situation of Figure 8-12 the uplifted parcel becomes increasingly warmer relative to its surroundings, increasing its tendency to rise.

When a rising air column reaches its dew point temperature, the adiabatic lapse rate changes dramatically. This change may produce unstable air above the dew point level while it remains stable below (Fig. 8-13). Although a 5°F difference exists at ground level between a heated parcel of air and its surroundings, that differential decreases as the heated parcel rises through a surrounding volume with a lapse rate of 3.8°. It is approaching stability, but before that happens it cools to its dew point. The adiabatic lapse rate now drops from 5.5°F per 1,000 feet down to 3.2°F. The lapse rate of the surrounding air is now higher than that of the rising air. The rising column becomes increasingly warmer, and lighter relative to its surroundings thus increasingly unstable.

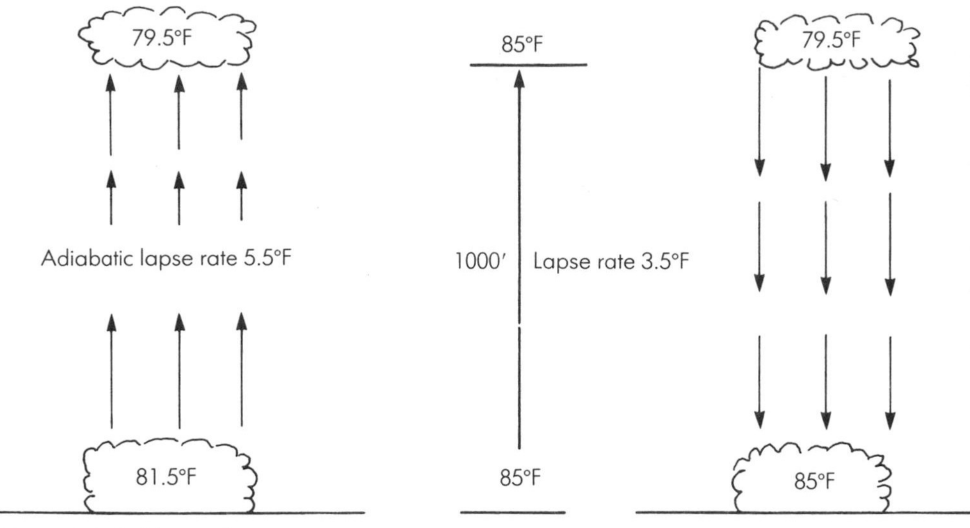

Figure 8-11. Stable condition.

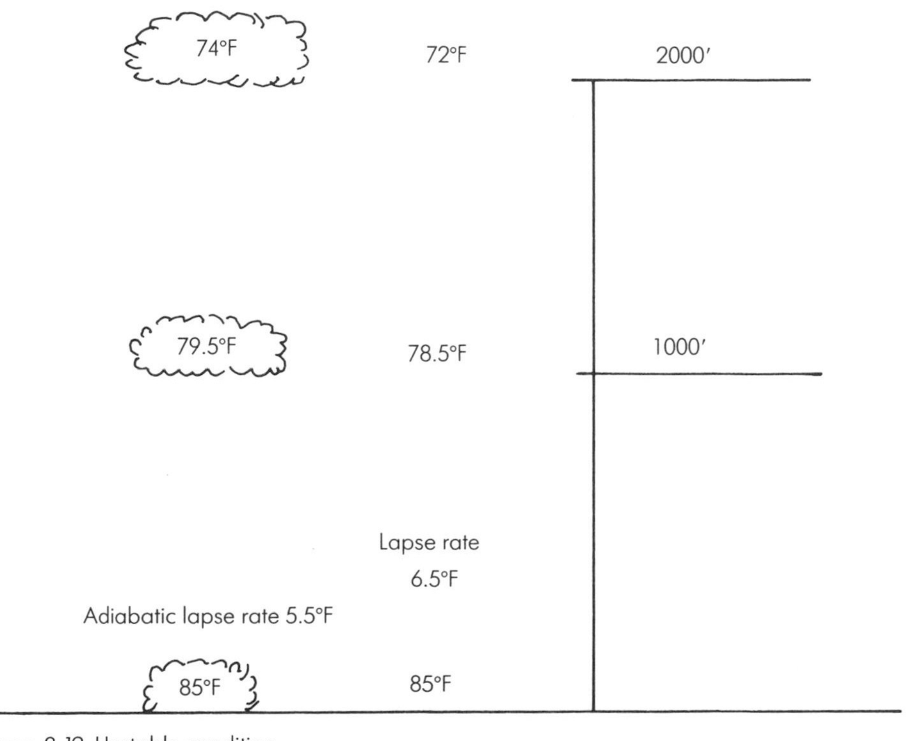

Figure 8-12. Unstable condition.

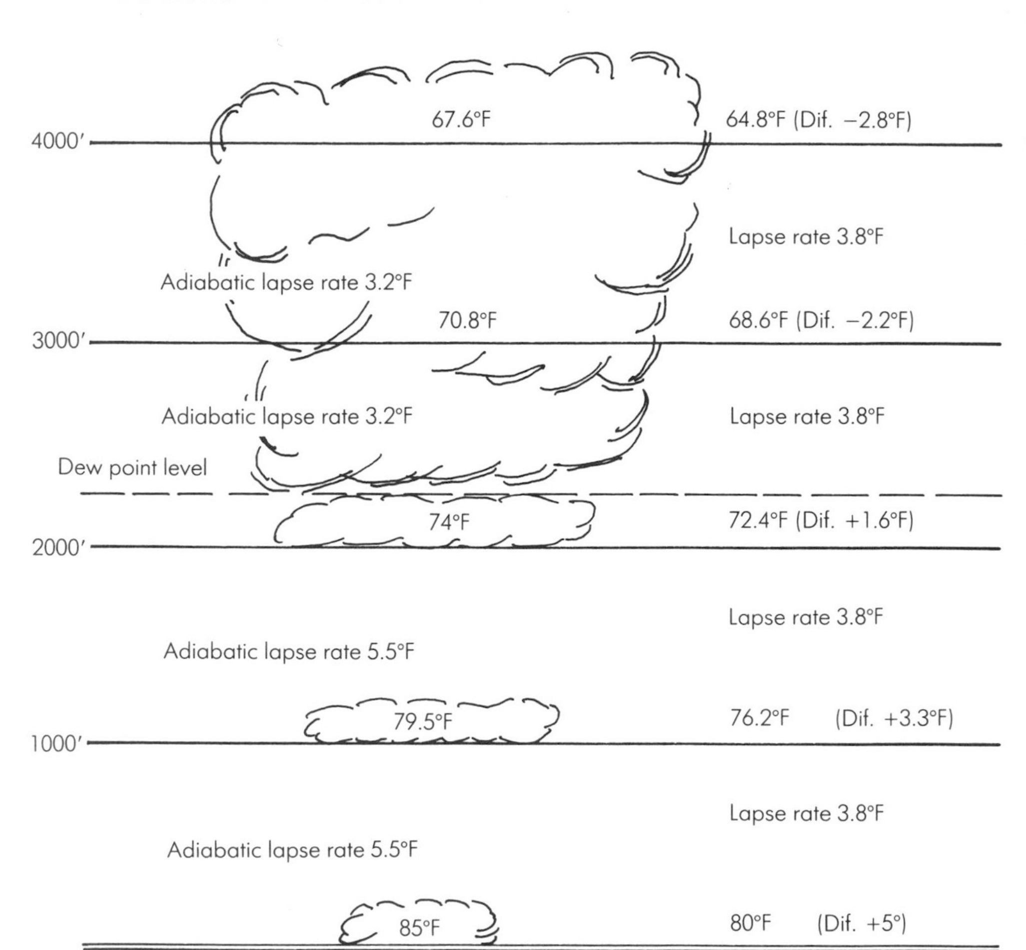

Figure 8-13. Conditionally unstable condition. Below dew point, level temperature differences are decreasing with altitude. Above that level they increase.

SUMMARY

Stability will exist when the lapse rate is less than the adiabatic rate (see Fig. 8-11). Vertical air movement will be minimal.

Instability will exist when the lapse rate is greater than the adiabatic rate (see Fig. 8-12). Considerable vertical air movement and turbulence will exist.

Conditional instability will exist when the lapse rate falls between the dry and the wet adiabatic rate. Air will be stable with limited vertical movement below the dew point level becoming unstable with strong vertical movement and turbulence above that level.

As we shall see, the sailor will find atmospheric stability, or the lack of it an extremely reliable and useful indicator regarding future weather changes. Without taking exhaustive temperature measurements, the mariner can deduce a great deal concerning atmospheric stability by simple observation. Smoke from a chimney, for example, that rises a short distance then floats off horizontally is an indicator of stability. Rapidly rising vertical smoke columns indicate unstable tendencies.

Stratus clouds (see Fig. 9-12) indicate stability and usually conditions such that weather changes will occur slowly. If the weather is deteriorating it will do so gradually—you've got some time! On the contrary, if the clouds are towering cumulus types (see Fig. 9-17) the air is very unstable. Weather conditions can worsen suddenly. In highly unstable conditions watch the sky and the wind, and be ready to run for shelter at flank speed if the weather starts to change.

Atmospheric stability also has a pronounced effect on winds. The winds in stable air tend to be steady in direction and velocity. Winds in an unstable atmosphere tend to be puffy and directionally unsteady.

ATMOSPHERIC PRESSURE

In addition to heat, and humidity (or water content) a third basic factor influencing atmospheric activity and resulting weather changes is atmospheric pressure. As mentioned earlier, the atmosphere of our planet decreases in density with altitude. The density at any specific altitude is simply the weight of the air above that level as measured by a pressure measuring instrument, the *barometer*.

The "bottom" of the atmosphere is at sea level. Here it is at maximum density and therefore maximum pressure as well. On the left side of Figure 8-1, the average pressures at various altitudes may be found below the word "Pressure." Average pressures at different altitudes are given in *millibars*. For simplicity these ascending pressures are shown as powers of 10 at the bottom of the column just above sea level 10^3 thus means 1,000. The average so called normal sea level pressure is 1,013.2 millibars, or just a trifle above the 1,000, which might be very interesting if one had any idea as to what is meant by a millibar.

THE BAROMETER

Fortunately, that's easy, but we have to go back a bit into history. In 1643 Evangelista Torricelli invented the mercury barometer (see Fig. 1-1) to measure atmospheric pressure. With this device, pressure was indicated by the height of a column of mercury in a glass tube. It was quickly found that this height varies from time to time, and from place to place. Therefore, as starting points for comparison it was necessary to define a standard atmosphere, and a normal pressure.

The standard condition finally adopted was a temperature of 15°C (59°F), at sea level, at latitude 45 degrees. Normal pressure for these conditions was a *bar*. Since a bar is just under 30 inches it was divided into thousandths, or millibars, for ease in observing and recording changes.

Then, it appears, a discrepancy crept in. The bar was set at 29.53 inches of mercury, but normal pressure turned out to be 29.92 inches. The result is that normal pressure turns out to be 1013.2 millibars, not an even 1000 millibars.

As discussed in Chapter 3, the early mercury barometer has been replaced for marine use by the aneroid barometer (see Fig. 3-1). This instrument is compact, and essentially unaffected by the movement of the boat. It can easily be permanently mounted on a cabin bulkhead for ready reference. It operates by converting changes in the pressure of the air on a small vacuum chamber into readings on a circular calibrated dial. Before installation, a barometer should be adjusted to match any standard barometer located nearby using the adjustment screw placed on the back for this purpose. Painstaking accuracy in setting a barometer is not important. Your interest is in the direction and magnitude of changes and the speed with which they are happening. Exact pressure is seldom critical for your purposes.

Normally there will also be an adjustable indicator needle set in the cover glass above the barometer needle. Periodically this indicator should be set to match the barometer needle, and the time logged. At regular intervals the barometer is then read and any change noted. A succession of readings over time reveals whether pressure is rising or falling, and how rapidly.

PRESSURE, TEMPERATURE, AND VOLUME OF AIR

The physical relationships between the temperature, the pressure, and the volume of parcel or a mass of air are fundamental to any understanding of weather phenomena. *Boyle's Law, Charles Law, and the Law of Gay-Lussac* explain these relationships.

The relationship explained by Boyle's Law is that pressure and volume are inversely proportional. If a quantity of gas occupying, for example, two cubic feet is compressed into a one cubic foot space (one half the original volume) its pressure will be doubled. If its original pressure were 15 psi, it will now be 30 psi. This assumes that the temperature remains unchanged.

Later, Charles Law showed that when the volume occupied by a gas is held constant, its pressure varies directly with temperature. If the temperature is raised the pressure will go up as well. Conversely, lower the temperature and pressure will drop. Also, if pressure is lowered, the temperature will drop as well. Let the air out of a tire with your hand in the airstream. It feels cold because its pressure has just dropped dramatically.

Boyle relates pressure and volume; Charles relates pressure and temperature, Gay-Lussac covers the other possible combination—volume and temperature. This time we find again a direct relationship. As temperature is increased, volume increases as well with pressure now constant.

Thus, as we have seen in conditions of atmospheric instability, when a quantity of air is heated it wants to expand—*Gay-Lussac*—(Fig. 8-7). As it expands to fill a larger cubic volume of space, its pressure decreases—*Boyle*. The decrease in pressure is accompanied by decreasing temperature—*Charles*. These relationships between the temperature, the pressure, and the volume of gases explain both local movements of small parcels of air, and the major atmospheric circulation patterns of large air masses.

ATMOSPHERIC PRESSURE VARIATIONS

As we have seen, temperatures vary moving up, or down, or horizontally through the atmosphere. We now find that in accordance with Charles Law pressure varies along with the temperature when volume is constant. As one increases so does the other. However, the atmosphere is not bottled up in a handy container, so as the pressure in a mass of air increases that mass expands in volume following Boyle's Law. As it expands it becomes less dense, thus lighter, and now rises. The laws of gases thus explain the causes of atmospheric stability and instability.

Looking back at the large scale pattern of the heating of the atmosphere (see Fig. 8-4) the most intense solar radiation strikes the equatorial areas. The least intense is found in the polar regions. According to the laws of gases this would lead one to expect to find warm, low pressure, rising and expanding air in the equatorial belt, and cold, high pressure, contracting and subsiding air near the poles. This is exactly what is found.

Ideally then, the overall pressure picture of the planet should show (Fig. 8-14) hot, low pressure air converging at the equatorial belt rising and expanding poleward at high altitude. It should then cool as it moves away from the equator arriving at the poles as cold, high pressure air which then descends and diverges from the poles back toward the equator.

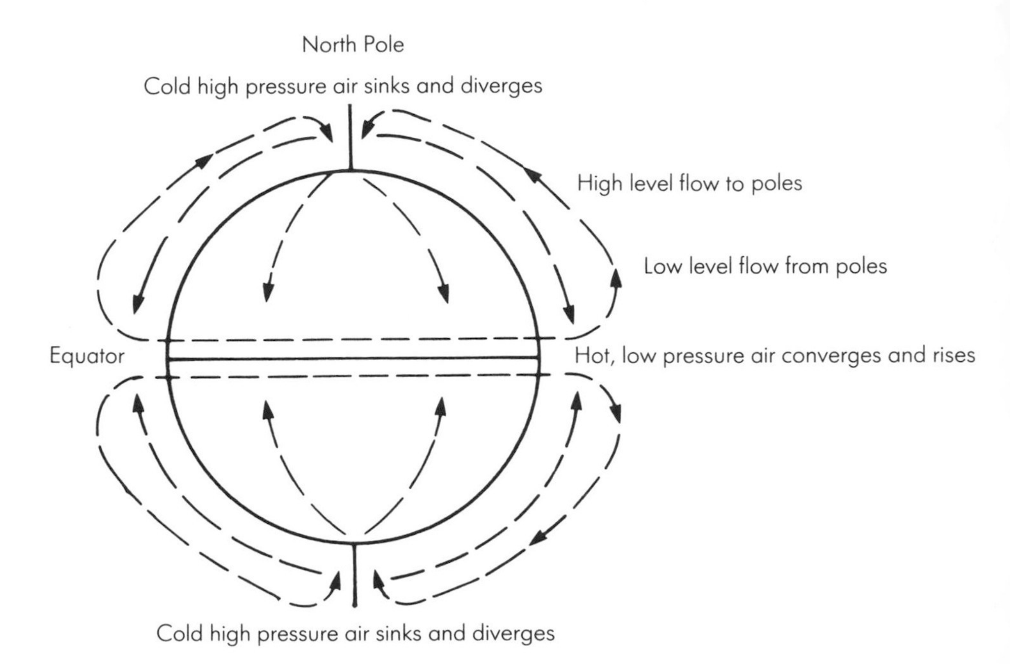

Figure 8-14. Idealized pressure and planetary air flow pattern.

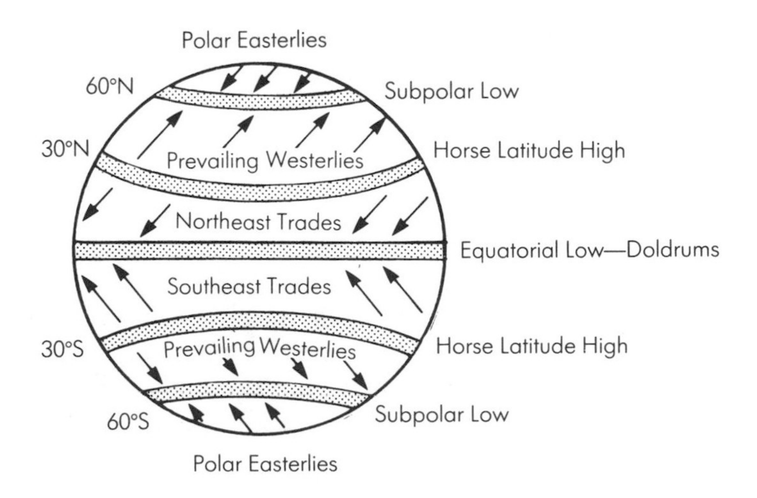

Figure 8-15. Primary wind and pressure systems of the earth.

Unfortunately, a combination of factors causes this lovely, simple picture to become more complicated. The actual planetary pressure system reveals two high-pressure belts, and two low-pressure belts between the equatorial low and the two polar highs. (Fig. 8-15) To see how this affects the sailor, we need to examine the major systems of horizontal air movement on the planet.

GENERAL PLANETARY AIR CIRCULATION

The air envelope of the earth is constantly in motion both vertically (up or down), and horizontally (the winds). The energy powering this movement is still the heat derived from the sun's radiation. That heat, as we have seen, is very unevenly distributed over the surface of the earth, and through the atmosphere as well.

The constant air movement reflects the endless attempts (doomed to failure) of the atmosphere to balance out air temperatures and pressures. The failure to accomplish this even spread of temperatures and pressures is inevitable. Additional, unevenly distributed heat energy is continually being added to the atmosphere by solar radiation, while heat is constantly being lost unevenly to space by albedo and infrared radiation.

The total amount of heat received by the entire planet is balanced by the total amount lost to space. However, both what comes, and what goes are unevenly distributed.

The winds produced by these uneven temperatures and pressures fall into two categories: primary wind systems, and secondary winds. The primary wind systems are the easterly trade winds of the tropics, the prevailing westerlies of the temperate zones, and the polar easterlies (Fig. 8-15).

Secondary winds are associated with the fronts dividing air masses (Chapter 10), large area cyclonic storm systems (Chapter 11), comparatively small area violent storms (Chapter 12), and many types of local atmospheric disturbances and irregularities. These all occur within the overall framework of the primary wind systems of the earth.

There is no need for the sailor or the fisherman to be concerned with the explanations, and the disagreements, of distinguished meteorologists as to why these primary wind patterns exist in exactly the way that they do. Consequently, we shall merely describe what has been observed, but first let us look at a mechanism that we shall encounter again and again in connection with winds. That mechanism is known as the *Coriolis effect*.

CORIOLIS EFFECT

The earth is approximately 21,710 nautical miles around at the equator. As one moves away from the equator the circumference of the earth at any parallel of latitude decreases as latitude increases (Fig. 8-16). Therefore, a place at latitude 30° travels a distinctly shorter distance to accomplish its one complete rotation per day than one at the equator. A point at latitude 60° moves a much shorter distance yet.

Since at each higher latitude it takes the same amount of time to travel increasingly shorter distances, the actual speed of surface movement is decreasing. A point on the equator moves at a speed of approximately 868 knots to complete its daily revolution while one at latitude 60° moves at only 434 knots.

Let us now suppose that some air at latitude 60°N starts to move south. It has a rotational velocity of 434 knots. As it goes south it moves over areas that have increasingly higher rotational velocity than it has. To an observer on the surface, it has been deflected and seems to come from east of north (Fig. 8-17).

Air moving up from the south, let's say the equator, has a rotational speed of 868 knots which is going to be faster than any place it passes over. To a ground observer it too is deflected appearing to come from southwest rather than south. In the Southern Hemisphere the deflections are reversed (Fig. 8-17).

PRIMARY WIND SYSTEMS

The laws of gases have shown us that as air is warmed it expands. Expansion decreases its pressure, and its density allowing it to rise. In the equatorial regions, which receive the greatest amount of solar heating of any area on the planet, this is exactly what happens. What amounts to a gigantic convection belt is formed where air is being heated. It expands, dropping in pressure, and rising to be replaced by air from both sides of the equator. This area is called the *Intertropical Convergence Zone*. The weather in this belt is hot and humid. Winds are extremely light and variable. Rain squalls are very frequent. The exact location of this belt moves north and south with the seasons.

Air from the north moves south, and air from the south moves north toward the low Convergence Zone in the middle. However, our new aquaintance, the Coriolis effect, now comes into play. Both incoming winds are deflected. The one from the north becomes northeasterly. The one from the south becomes southeasterly. Because these winds are extremely steady, they were used whenever possible

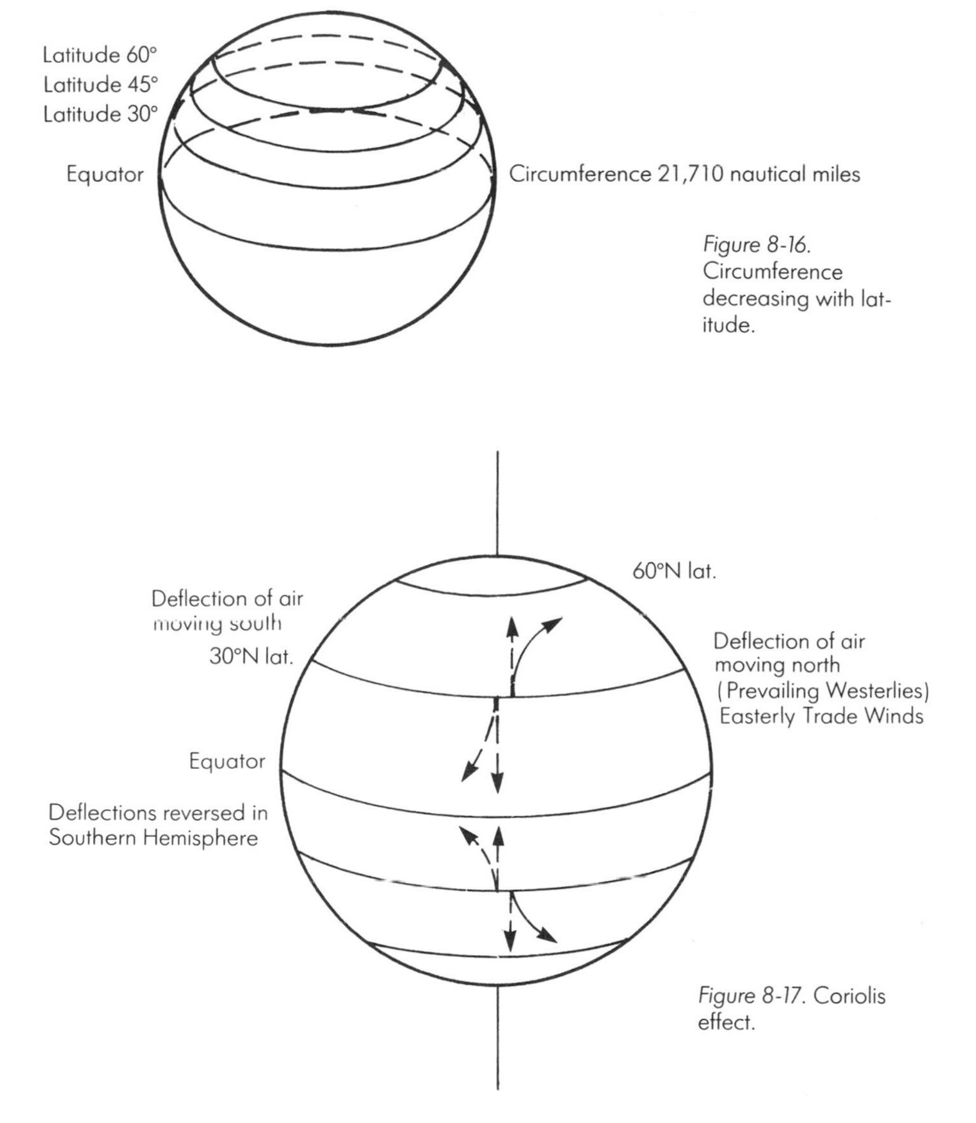

Figure 8-16. Circumference decreasing with latitude.

Figure 8-17. Coriolis effect.

by the old square rigged commercial sailing ships, and hence became known as the "trade winds" (see Fig. 8-15).

Meanwhile, the air that has been convected upward is moving away from the equatorial belt at high altitude toward both north and south. These high altitude winds are also deflected by the Coriolis effect causing them to blow substantially opposite to the surface trade winds.

At approximately latitude 30° (N and S) a band of high pressure subsiding air exists. Here, as at the equator, winds are light and variable, but the weather is totally different. It is generally clear with very little rainfall. These are termed the "horse latitudes," a term which also goes back to the days of square rigged sailing ships. Such ships often were stalled for many days in the light winds of these latitudes. When the Spanish were transporting horses across the Atlantic for use in the Americas, they were often delayed in this way. Horses drink a great deal of water. During extended delays they ran short of water and the horses had to be thrown overboard giving rise to the name.

The descending air of these belts diverges both north and south. In the Northern Hemisphere, the winds blowing south are diverted by the Coriolis effect to become the tropical easterly trade winds blowing toward the low pressure equatorial doldrums. The winds blowing toward the north are also diverted by the same force to become the prevailing westerlies of the temperate zone.

The prevailing westerlies blow in the same direction as the rotation of the earth. At and close to surface level they tend to blow more strongly than the trade winds of the tropics. The strongest winds at sea level in the westerly wind belt of the northern hemisphere are found at about latitude 45°N.

The middle latitudes of the prevailing westerlies, where most of our boats are operated, are characterized by strong temperature contrasts which produce enormous amounts of potential energy. Large air masses with contrasting characteristics absorb this energy giving rise to the major cyclonic and anticyclonic weather systems that constantly move through these latitudes.

The prevailing westerlies blow from the high pressure band at the horse latitudes toward another band of low pressure, the subpolar low, at about latitude 60° (N and S). The northern subpolar low is constantly pushed back and forth by the large cyclonic weather systems moving through the middle latitudes to its south.

The major continental land masses are concentrated in the Northern Hemisphere. The resulting contrasts in heating and cooling between land and sea, cause the weather systems in the middle latitudes here to be particularly turbulent.

Winds in the polar areas again become easterly. Most of us will probably be just as happy to find that bit of information totally useless; we have no intention of going boating in polar waters.

Superimposed on these primary wind systems are a wide variety of secondary winds sometimes reinforcing, and at others countering the primary winds. The secondary winds having the most widespread weather influence are those associated with major air masses, and the frontal disturbances that occur where they meet.

9

CLOUDS

In the absence of instruments or gauges of any kind, the sailor who has trained himself to read the sky, and its passing cloud patterns, can still predict the weather for anywhere from a few hours to several days in advance. How far in advance the clouds will indicate coming weather patterns varies with the type and size of the advancing weather system.

Thunderstorms give brief warning, but are over quickly. A major cyclonic disturbance gives ample advance warning, but is likely to take several days to pass. There is much truth in the old saying quoted in Chapter 1:

> Long foretold, long last,
> Short notice, soon past.

CLOUD FORMATION

Clouds, as you well know, are of many different shapes and types. The various cloud types reflect the weather conditions under which they were formed. Thus, by reading the passing cloud types, their direction and speed of movement the sailor may obtain an insight into the weather activity in his vicinity.

As we saw in Chapter 8, the atmosphere is made up of fixed proportions of oxygen, nitrogen, and several other gases, varying amounts of natural, industrial, and machine-made dust and pollutants, plus varying amounts of water vapor.

The amount of water a particular parcel of air can hold in vapor form depends on its temperature. As air becomes warmer its capacity to hold water vapor increases. As it cools that capacity decreases (Table 9-1).

ABSOLUTE HUMIDITY VALUES FOR SATURATED AIR

TEMPERATURE, °F	WATER VAPOR, *grains per cu ft*	TEMPERATURE	WATER VAPOR, *grains per cu ft*
0	0.479	50	4.108
5	0.613	55	4.891
10	0.780	60	5.800
15	0.988	65	6.852
20	1.244	70	8.066
25	1.558	75	9.460
30	1.942	80	11.056
35	2.375	85	12.878
40	2.863	90	14.951
45	3.436	95	17.305

FROM: William L. Donn, *Meteorology*, New York: McGraw-Hill, 1965.

When a volume of air at a specific temperature has picked up as much water in vapor form as it can hold it is saturated. It is at 100 percent relative humidity. Any decrease in temperature will result in condensation of some of its vapor into minute water droplets which, suspended in the air, make up clouds.

We saw in Chapter 8 that, while sometimes interrupted by inversions, air temperature normally decreases steadily with altitude in the lower atmosphere. The rate of that decrease (the lapse rate) varies from day to day, and from place to place. It also changes abruptly at the altitude where condensation starts due to the release of the latent heat that was contained in the condensing water vapor.

FORMATION OF CUMULUS CLOUDS

As we know, an area of the earth's surface warmed by the sun heats the air immediately above it by a combination of direct conduction and radiation. The warming of that air causes expansion. As a parcel of air expands it becomes less dense, thus lighter than the air around it. It may now become detached from the ground layer and convectively start to rise (Fig. 8-7).

It continues to expand as it rises, and also starts to cool at the dry adiabatic rate of 5.5°F per 1,000 feet. At some level, which depends on its moisture content, it becomes saturated. It is now at 100 percent relative humidity (also called its dew point, remember?) and contains all the water vapor it can handle. As it moves above that level and continues to cool, condensation takes place. Minute water droplets form, making the rising air parcel now visible (Fig. 9-1) as a cumulus cloud.

Once condensation has started, the cooling rate of the rising air changes to the wet adiabatic rate. As we saw in Chapter 8, the differential between the adiabatic rate and the lapse rate of the surrounding air determines the altitude at which the rising air stablizes.

An elaborate system of Latin nomenclature is used to describe the various kinds of cumulus clouds and other cloud types as well. In 1803 a London pharmacist named Luke Howard devised the basic nomenclature system which has since been adopted by the World Meteorological Organization (WMO). For our purposes we need to

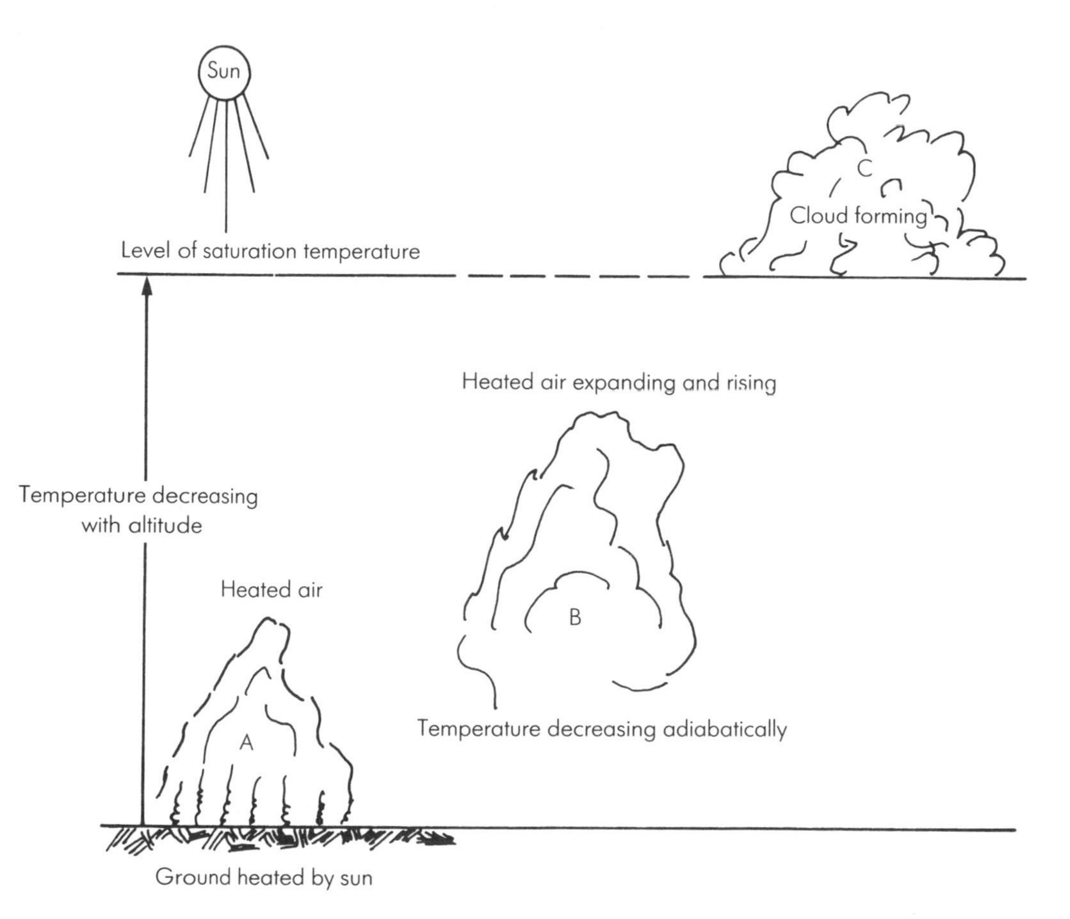

Figure 9-1. Cumulus cloud formation.

deal with only a few of the many variations, and be of good cheer, we'll get away with very little Latin.

Cumulus clouds vary widely in the heights at which they form, the heights of the clouds themselves, their speed of formation, the area of the sky they cover, and the length of time they persist after formation. Cumulus clouds are also called *convective* clouds since they are produced by air rising due to convection. The rising air column that culminates in a cumulus cloud is also called a *convective cell.*

Very large and active convective cells develop into thunderstorms (to be discussed in detail in Chapter 12); however, the clouds from which they develop are of the cumulus type (Fig. 9-2). The cumulus clouds that become thunderstorms are particularly tall clouds. Their bases are at heights between about 4,000 and 5,000 feet with tops reaching up between 40,000 and 50,000 feet.

Cumulus clouds form in unstable air which is air easily put into vertical movement because its lapse rate at the time exceeds the adiabatic rate. When cumulus clouds are forming in unstable air above which lies a layer of stable air (Fig. 9-3) the cumulus clouds flatten out when they reach the stable layer. They then change into a cloud type transitional between cumulus and stratus clouds.

Due to the way in which they are formed, the different types of cumulus clouds all appear rounded and bulbous. However, before discussing cumulus cloud variations in more detail let us look further at the mechanics of cloud formation. We have seen that cumulus clouds are formed by air in vertical, or convective, movement. Another major cloud group is developed differently.

FORMATION OF STRATUS CLOUDS

The rounded shapes of cumulus clouds that result from the vertical upward boiling of heated air parcels differs sharply from the comparatively flat, featureless look of stratus clouds. The difference is due to the fact that stratus clouds form in air that is moving horizontally rather than vertically.

Just like cumulus clouds, the stratus type forms when a volume of air is cooled below its saturation (or dew point) temperature, thus causing its water vapor to condense out as minute water droplets.

As we shall see in Chapter 10, the major weather disturbances that move around our planet are caused by collisions between air masses having different characteristics. One common type of collision occurs when a volume of warm air meets a mass of cool or cold air. When this happens (Fig. 9-4) the warm air, being lighter, gradually slides up over the denser cold air.

A. Low cumulus fair weather "puff ball"

B. Cumulus with growing vertical developing

C. Tall cumulus with typical thunderhead "anvil" top forming

Figure 9-2. Cumulus developing vertically in unstable air—thunderhead forming.

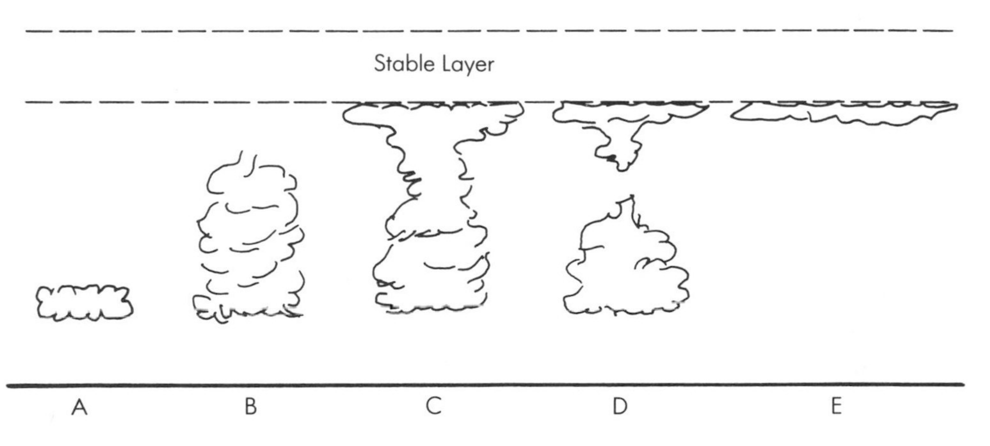

Figure 9-3. Cumulus cloud rising and spreading out under stable air layer.

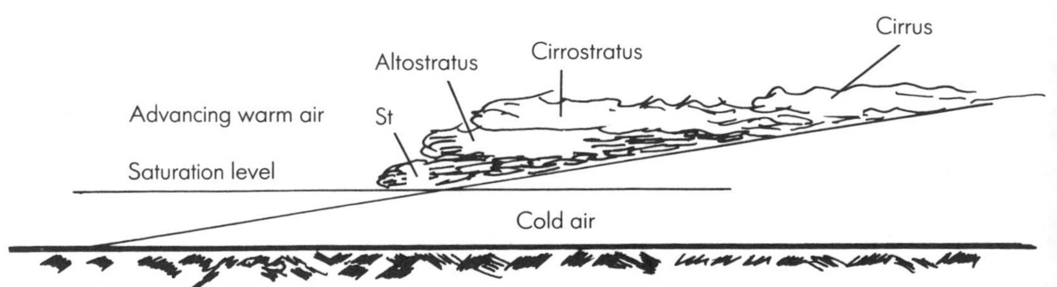

Figure 9-4. Cirrus—thin, wispy, ice crystal clouds. Cirrostratus—high, thin, translucent sheet cloud. Altostratus—middle height, flat gray, featureless sheet cloud. St—stratus—low, thick, flat, may be patchy, may drop rain or snow.

Cooling as it rises, the warmer air eventually reaches its saturation temperature. This starts the formation of large, flat sheets of stratus clouds. The altitudes at which stratus clouds form and the densities of the cloud layers vary considerably. The thickest stratus clouds are at low altitudes. Stratus layers then decrease in density with height, becoming thin translucent sheets at high altitudes.

CLOUD DISSIPATION

Clouds are formed by the cooling of a volume of air below its saturation temperature resulting in condensation. They are dissipated when that air warms again because warming increases its capacity to hold water in vapor form. As a volume of air containing cloud is warmed, its suspended water droplets gradually evaporate disappearing into water vapor.

The warming necessary to cause cloud dissipation, may result from mixing of clouds with rising or surrounding warm air. It may also be produced by conditions that cause the air containing clouds to move downward (Fig. 9-5). This is called *subsidance*. Descending air warms (increasing its capacity to hold water vapor) at the same adiabatic rate that rising air cools (5.5°F per 1,000 feet) Steadily descending air continuously increases in its capacity to take up water vapor resulting in rapid cloud dissipation.

On a clear, sunny day one often sees the typical fair-weather cumulus "puff balls" forming over a local convective updraft only to move on to dissipate over a nearby downdraft. The generally clear skies of the horse latitudes are a product of largescale subsidance. Other conditions leading to largescale subsidance will be discussed further in connection with the movements of air masses in Chapter 10.

GREENHOUSE EFFECT

Another phenomenon that has considerable influence on cloud dissipation is the *greenhouse effect* (Fig. 9-6). This was briefly mentioned in Chapter 8. The heat which powers the weather systems of our planet comes from solar radiation. This radiation reaches earth in the form of relatively short wavelength visible light. Even during overcast weather a great deal of that light, diffused to varying degrees, gets through to the surface.

When this radiation strikes the earth's surface part of it is immediately reflected away as visible light—albedo. Another portion is absorbed as heat by that surface. A third fraction is transformed from

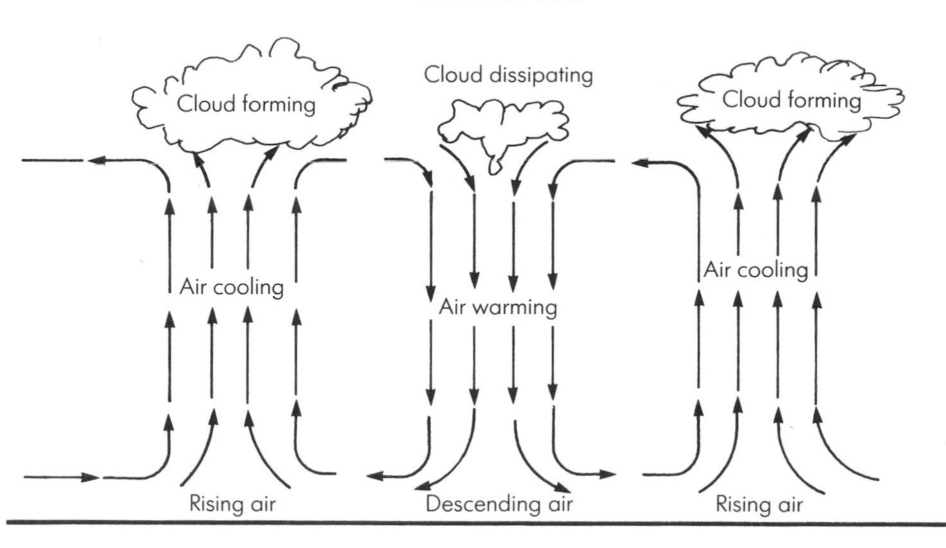

Figure 9-5. Clouds forming in rising air—dissipating in descending air.

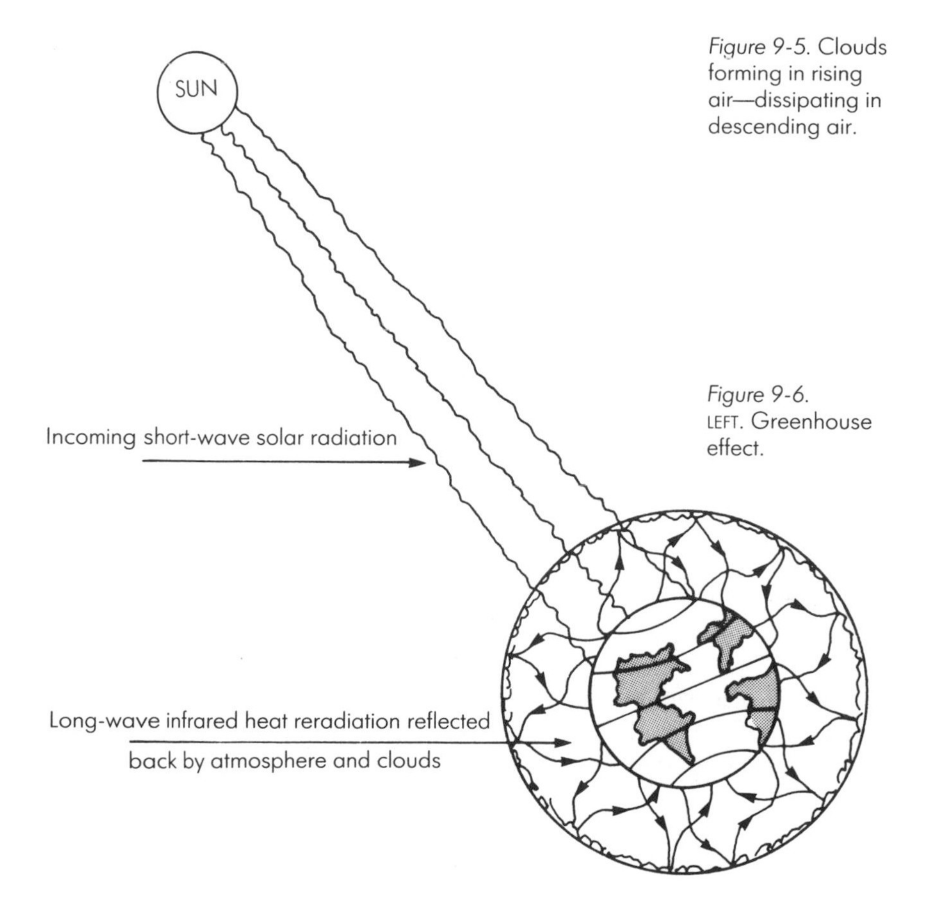

Figure 9-6. LEFT. Greenhouse effect.

short wavelength visible light into long wave invisible infrared heat radiation.

The portion reradiated in the long wave infrared heat range has very different characteristics from the original short wavelength light radiation. The short wavelengths in the visible spectrum are partly blocked—or actually absorbed—by an overcast sky (see Fig. 8-6). However, a great deal passes through. The portion of the short wavelengths that is transformed and reradiated as infrared heat is, however, almost totally blocked by the overcast. Part of it is absorbed in the cloud deck, and part is reflected back to warm the space below.

As the bottom of the cloud deck, and the area below it, are warmed by infrared radiation the water vapor capacity of that air is increased. Water droplets from the clouds then evaporate, and the cloud partially or totally dissipates from the bottom up.

You have certainly often seen overcast days when in the early morning the cloud layer was very thick, totally obscuring the sun. Later the same day the clouds thined enough that a watery sun became visible, only to thicken again totally obscuring the sun in the late afternoon. This is commonly the result of heat supplied by infrared radiation from below.

This is called the "greenhouse effect" because a greenhouse is warmed in the same way. Incoming visible light passes readily through the glass, but the portion transformed inside the building into infrared radiation does not pass back out through the glass. As we saw in Chapter 5, this same phenomenon enables us to predict that it will not be nearly as cold on cloudy nights as on clear ones. On a clear night infrared heat is rapidly radiated to outer space. On a cloudy one the cloud layer reflects much of it back.

CLOUD CLASSIFICATION

Back in 1803, when he set about to classify cloud types, Mr. Luke Howard, being a pharmacist, was familiar with the use in scientific fields of Latin names for the classification of plants and animals. Since this was also to be a scientific classification project, in his mind it obviously had to be done in Latin as well. The result was ten basic cloud types along with a number of variations of those types, all in Latin!

We have already met two of these Latin terms: cumulus and stratus, and have seen the basic difference in the way these cloud types are formed. Depending on differing conditions of temperature and humidity, both types may form at various altitudes. The altitudes at which we see them, and the sequence in which they pass our location tell us much about approaching weather systems.

Howard's ten basic cloud classifications as now found in the International Cloud Atlas are:

NAME		TYPICAL HEIGHT	
Cirrus	Ci	High	16,500′ to 45,000′
Cirrostratus	Cs		
Cirrocumulus	Cc		
Altostratus	As	Middle	6,500′ to 23,000′
Altocumulus	Ac		
Stratus	St	Low	Under 6,500′
Stratocumulus	Sc		
Nimbostratus	Ns		
Cumulus	Cu	Height varies with vertical development	
Cumulonimbus	Cb		

Ok, that's it for the Latin! Now let's see what it means.

HIGH CLOUDS

CIRRUS (Fig. 9-7)

Cirrus are very high, very thin streaky clouds. They have a delicate, fibrous appearance without shading, and are generally white. The temperature at their altitude is well below freezing, consequently they are composed of ice crystals. They are thin because air that cold has little water vapor left.

Cirrus appear in varied forms. Sometimes they are isolated tufts, sometimes long, thin lines. They may branch in feathery plumes, or they may run in parallel bands across the sky seeming to converge toward the horizon. The "mare's tails" of Figure 9-7a may become more abundant and thicker as in 9-7b. They may then spread out over the sky in bands as in 9-7c. In advance of a cyclonic disturbance those bands may indicate the direction of the low pressure or storm area.

Cirrus are by no means certain indicators of coming bad weather. They are more generally associated with fair weather. However, when a storm is on the way they are often the sailor's first visible sign.

Figure 9-7.

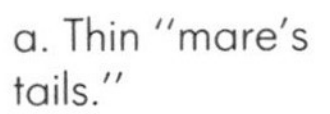

a. Thin "mare's tails."

b. More abundant cirrus clouds.

c. Banded cirrus clouds.

CIRROSTRATUS (Fig. 9-8)

When the cirrus clouds have spread over the sky to form a thin, white filmy layer they have become cirrostratus. The sky now presents a milky or veiled appearance. Haloes often appear around the sun or the moon. The thinness of these clouds are indicative of their great altitude which also accounts for their composition of ice crystals. Like other stratus clouds, they were formed by air moving horizontally and rising gradually over colder lower layers (see Fig. 9-4). If these clouds continue to thicken and lower expect rain within 24 hours.

Figure 9-8. Cirrostratus layer developed from earlier cirrus.

CIRROCUMULUS (Fig. 9-9)

These clouds appear as small, white flaky or scaly bits very close together, and covering quite large areas of the sky. They are without shadows. Usually they seem to be arranged in bands or rows like fish scales causing sailors to call them "mackerel sky." Although called a form of cumulus, these clouds are formed not by rising convective air currents, but by the degeneration of cirrus or cirrostratus clouds. Being another of the very high cloud types, cirrocumulus are also composed of ice crystals. Like cirrus they are usually associated with fair weather, but may precede a storm if they turn gray, thicken, and lower.

Figure 9-9. Cirrocumulus clouds developing from cirrus.

MIDDLE ALTITUDE CLOUDS

ALTOSTRATUS (Fig. 9-10)

Dropping down from high level clouds to middle level types, altostratus are uniform, flat, pale gray or gray-blue sheets covering all or large sections of the sky. Sometimes they occur as wide bands. The sun may be totally obscured, or it may be visible as a weak, watery light spot in an otherwise featureless sky. A complete or nearly complete absence of shadows is typical of the diffused light below altostratus clouds.

Their thickness varies depending on the altitude at which they form. Remember, the higher you go the colder the air becomes, and the less water vapor it has left. Thus, the higher, altostratus clouds are when they form, the less moisture there is available, resulting in thin clouds. The lower they are when forming the more moisture is available, and the thicker they become. At middle and higher latitudes considerable rain or snow falls from lower level altostratus type clouds.

ALTOCUMULUS (Fig. 9-11)

These clouds form as rather bulbous elliptical shapes sometimes occurring individually, but more often in groups. When grouped they may form as confused masses, or in fairly definite bands with alternating bands of clear sky between. The banded form sometimes appears briefly in advance of a thunderstorm, or may thicken and lower bringing showers. However, they do not bring prolonged bad weather. The coloring varies from pure white through a wide range to very dark gray.

Individual altocumulus clouds frequently have an elongated lenticular, elliptical shape easily distinguishable from the low level cumulus cloud. It has a higher base altitude, and it lacks the vertical development commonly found, as we shall see, in the lower type cumulus.

Figure 9-10. Altostratus clouds. High, thin clouds. Sun shows as a fairly distinct disk.

Figure 9-11. Altocumulus clouds.

LOW CLOUDS

Figure 9-12. Stratus clouds. Sun showing through cloud layer as an indistinct light area.

STRATUS (Fig. 9-12)

This is a low, uniform gray sheet or layer. It shows no definite form or structure. While it generally covers the sky completely, it may be partly broken by long openings. Low stratus is often impossible to distinguish from a high fog. It often develops from fog. The bottom level of the fog may dissipate through the greenhouse effect, leaving low stratus, or high fog, whichever you prefer to call it.

In the presence of stratus clouds, surface winds are usually light. Increasing wind will generally break the uniform stratus layer into shreds, and often disperse it entirely.

Stratus clouds are normally evidence of weak weather activity. They bring dull, dreary conditions rather than stormy conditions.

STRATOCUMULUS (Fig. 9-13)

These clouds form when rising cumulus convection cells reach a stable layer in the atmosphere (see Fig. 9-3). They then spread out into soft, gray roll-shaped masses. Often they appear as long, gray, parallel bands covering all or most of the sky.

Stratocumulus vary considerably in altitude since they form at whatever level a stable layer happens to exist. They form during the day when convection cells are actively forming cumulus clouds, and are usually followed by clear skies at night.

***NIMBOSTRATUS** (Fig. 9-14)*

These are thick, dark gray, shapeless, rain clouds. At the end of a succession of gradually thickening and lowering clouds the nimbostratus move in and a steady protracted rain or snow follows. They do not produce showers, but rather steady, drenching rains lasting anywhere from hours to days.

Winds may be light or strong, but normally steady rather than squally. There may be breaks in the rain, or the winds may die down and then return, but changes will be gradual not sudden. The clearing that comes after a spell of nimbostratus rain will also be gradual.

Figure 9-13. Stratocumulus clouds.

Figure 9-14. Nimbostratus clouds. Thick, dark, vague shapes, low. Steady long-lasting rain will soon start.

CUMULUS (Fig. 9-15)

As we saw in the discussion of cloud formation, cumulus clouds form with flat bases at the dew point altitude which is quite low. They are dense because the air in which they form carries considerable water vapor.

Because of the way they form, cumulus clouds typically show some *vertical development*. This vertical development may be slight, as in Figure 9-15a, moderate, as in 9-15b, or pronounced, as in 9-15c. The amount of vertical development in cumulus clouds is a clear indicator of the extent of atmospheric instability present. The very low, small

Figure 9-15a,b. HERE AND OPPOSITE. Cumulus—increasing vertical development.

patches of Figure 9-15a are sometimes referred to as fair-weather puff-balls.

Cumulus clouds are patchy and do not cover the entire sky, although they often appear to be thicker toward the horizon. This is an optical illusion as illustrated in Figure 9-16. As vertical development increases so do color contrasts, ranging from dazzling pure white through varying shades down to very dark grays (Fig. 9-17).

Cumulus clouds of low to moderate vertical development are fair-weather clouds. Their appearance after a stormy period heralds the return of clear, cool weather that will start out windy as well.

Figure 9-15c.

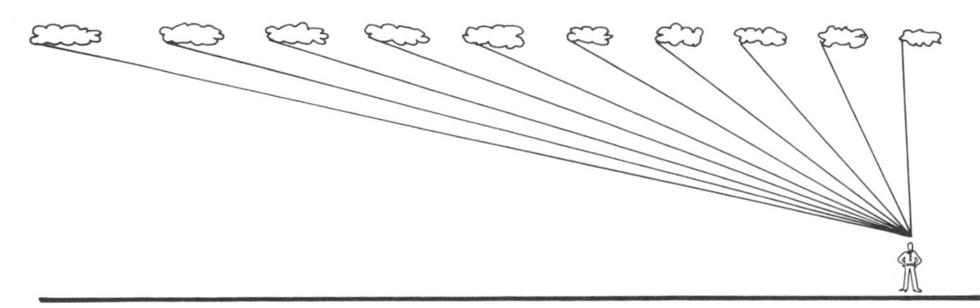

Figure 9-16. Cumulus clouds appear to thicken toward the horizon, only an optical illusion.

Figure 9-17. Strongly developed cumulus clouds show wide range of color gradation.

Figure 9-18. Rapidly forming cumulonimbus cloud.

CUMULONIMBUS (Fig. 9-18)

These clouds are formed over very vigorous and powerful convection cells which produce tremendously thick, high vertical development. They form and develop extremely rapidly in very unstable air into localized, often quite violent, thunderstorms which will be discussed in detail in Chapter 12. The cloud shown in Figure 9-18 became a full grown thunderhead in no more than fifteen minutes from the time the picture was taken.

SOME COMMON SKY CONDITIONS AND WHAT THEY MAY INDICATE

The passing cloud trains will reveal to you a great deal about the weather and possible future changes without the use of any instruments at all providing:

1. You keep an intermittent, but constant, watch on the clouds, keeping track of their direction of movement, and changes over time in types, thickness, and amount of sky coverage.

2. Wind direction and force. You don't need an expensive anemometer to tell you that the wind has increased, decreased, changed direction, or stayed about the same. Absolute velocity you don't need—you are watching for *changes*.

3. Temperature. You again do not need a thermometer to tell you it is warming or cooling. Is it about average for the season and time of day, or is it unusual?

These three factors are interrelated: the appearance of the sky, the wind, and the temperature. Watching these three factors over time will provide you with excellent clues to future changes.

CIRRUS—SCATTERED "MARE'S TAILS," SMALL CUMULUS BELOW

When high cirrus appear with cumulus puff balls below, or no other clouds below, look at the direction of cloud movement and feel the surface wind. When the surface wind and the cirrus are moving in the same direction, all is well. Weather change is unlikely.

If the cirrus is moving from left to right, and the wind is at your back, its direction sharply different from the direction of cloud movement, changes probably *are* coming. The cirrus may thicken to cirrostratus, then lower to altostratus, then to stratus, and probably nimbostratus. Wind will increase, temperature will drop, and it will rain. The change cycle may take anywhere from 6 to 12 hours or more. The slower the change, the longer the bad weather will last. When changes start, if they are happening quickly, batten down and head for port. If they are coming slowly, you can stay out and fish or sail for a while longer, but don't wait too long.

Figure 9-19. Sun showing as a fuzzy light area through a layer of altostratus clouds.

ALTOSTRATUS WITH HAZY SUN SHOWING (Fig. 9-19)

If this is what you see in the morning, it may have been preceded by cirrus and cirrostratus that you didn't see because they passed through during the night. If the wind backs and increases, and the clouds thicken, obscuring that hazy sun, rain is due in a few hours. If you're in port, stay there. If you're not, think about heading back, or getting out the foul weather gear.

CUMULUS—THICK, HEAVY, STRONG VERTICAL DEVELOPMENT (Fig. 9-20)

As the day has progressed the cumulus clouds have been getting thicker, darker, and taller. Squally, puffy winds and showers are likely very soon. Break out the foul weather gear and get ready to shorten sail and batten down in a hurry.

STRATUS—LOW, COVERING MOST OR ALL OF SKY (Fig. 9-21)

This sky indicates generally static conditions. There may or may not be light rain or drizzle. Winds will change little in velocity or direction. Temperature will also change little. It's not a pretty day, but it's not suddenly going to get a lot worse, or a lot better either!

Figure 9-20. Thick, heavy, tall cumulus in unstable air.

Figure 9-21. Low stratus cloud covering most of the sky.

Figure 9-22. Mixed cumulus and altocumulus as sky clears in the wake of a storm—the cold front has passed.

Figure 9-23. Fully developed thunderstorm.

MIXED ALTOCUMULUS AND CUMULUS—IMPROVING WEATHER (Fig. 9-22)
Temperature has dropped abruptly. It was raining, but it has now stopped. Wind has shifted from southwest to northwest. This broken sky precedes clearing, and cooling. Brisk wind will follow the rain and then decrease.

FULLY DEVELOPED CUMULONIMBUS THUNDERHEAD (Fig. 9-23)
This, from a distance, is a full grown thunderstorm. If it is heading toward you, or will pass close by, prepare for strong, squally winds and heavy rain. They will not last long, but while they do it will be rough going! This one may be far enough away that it will dissipate by the time it reaches your location, but, as we shall see in Chapter 12, these storms often form in groups.

If this storm is moving so as to pass well clear of you, fine, but watch out for its brothers and sisters anyway. Also note, if you are under sail, that while there are very strong winds within these storms they also tend to iron out the winds for some distance around. Expect calms before and after they pass.

10

AIR MASSES AND WEATHER FRONTS

As we have seen, temperature, pressure, and humidity in the atmosphere vary considerably at different times and places. We also know that the cause of these variations is rooted in disparities in the way solar radiation is received and absorbed by the earth's surface. In addition, and most importantly, we know that these differences in absorption of solar energy produce constant atmospheric turmoil. The atmosphere can never be at rest because temperature, pressure, and humidity constantly change as the movements of the planet vary the distribution solar energy.

The largescale atmospheric circulation over major areas of the earth's surface follows the basic pattern shown in Figure 8-16. Most of us will do most of our sailing in the latitudes of the prevailing westerlies. A few lucky ones will get down into the tropical belt of the easterly trade winds as well. Fewer yet will venture far enough north to feel the polar easterlies.

The prevailing winds in the belts from about latitude 30°N and S to about latitude 60°N and S are westerly. However, the prevailing winds, and weather conditions, in these belts are frequently disrupted by the movements of large air masses originating in either subpolar or subtropical areas. The passage of one of these large air masses is likely to drastically alter the weather in your vicinity for several days or up to a week at a time. Therefore, the sailor operating in the area of the prevailing westerlies should be able to recognize the signs indicating the comings and goings of such masses.

AIR MASSES

While the high and low pressure belts shown in Figure 8-16 are basically constant, they contain many internal variations caused by unevenness in the heating of the earth's surface. In consequence, large masses of air periodically stall and accumulate over certain fixed geographical areas. Such stationary air then tends to take on temperature and humidity characteristics typical of the area where it has stopped. When that occurs the stalled air then constitutes what is termed an *air mass.*

An air mass, thus, is defined as a large and distinctive body of air that has formed within the atmosphere. It is characterized by a *horizontal uniformity* of temperature and humidity. This is the result of its having lingered for some time over a homogeneous geographical region. "Large" means exactly that. An air mass covers hundreds of thousands of square miles in area, and extends many thousands of feet in altitude as well.

With the exception of occasional inversions, air temperature decreases with increasing altitude. Air masses are no exception to this general rule. The uniformity of temperature and humidity within an air mass is horizontal (Fig. 10-1). This means that at any place within the volume of space occupied by an air mass, whether at ground level, or at a height of 1,000 feet or at 5,000 feet, the temperature and humidity readings will be similar to those found at any other place within that mass at the same altitudes.

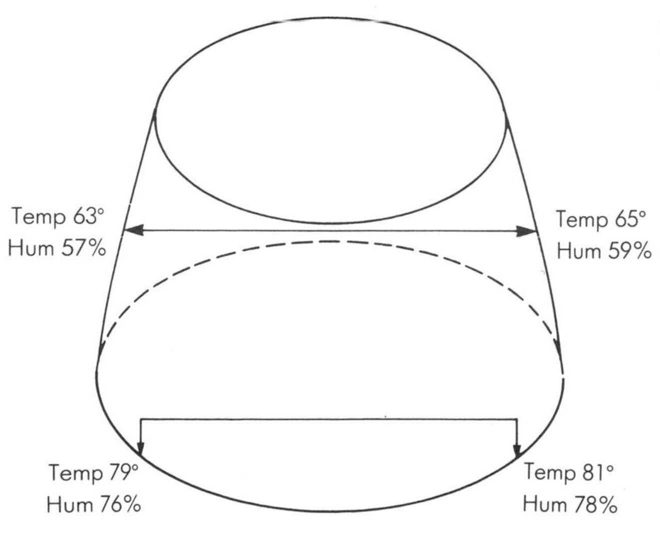

Figure 10-1. Temperature and humidity are substantially uniform *horizontally* in an air mass.

TYPES OF AIR MASSES

The geographical locations where large volumes of air can stall long enough to develop the horizontal uniformity typical of air masses are either completely over oceans, or completely over land. An air mass that forms over an ocean area is called a *maritime (m)* air mass. One that forms over land is termed *continental (c)*.

Most of the air masses that affect North America have sources either along the polar front to the north of the belt of prevailing westerlies, or in the subtropical calms of the "horse latitudes" to the south.

Those that form to the north are identified as *polar (P)*. The ones that form to the south are identified as *tropical (T)*.

The air masses that influence the weather in our sailing waters are generally going to be one of the following three types:

maritime polar	mP
continental polar	cP
maritime tropical	mT

Occasionally an *arctic (A)* air mass or an *equatorial (E)* air mass may invade our waters, but such invasions are comparatively rare.

Another type of air mass called *continental tropical (cT)* affects the Mediterranean Sea from time to time. However, the source region for this type of air is the North African Sahara desert. Other source regions for continental tropical air are the Arabian Desert, the Kalahari Desert, and the desert area of the Australian interior. Air masses of this type do not reach the coastal waters of North America (Fig. 10-2).

AIR MASS SOURCE REGIONS

MARITIME POLAR (mP)

In winter, the northeastern Atlantic and the northeastern Pacific are the primary sources of these air masses. The geographical area of these regions is rather small compared to some of the other source regions that will be discussed. In addition, the wind circulation in them is normally strong. Consequently, these air masses seldom tarry long enough to fully assimilate the typical properties of their source regions.

In summer, the maritime polar source regions move somewhat north of their winter positions, and are fed by a general outflow of

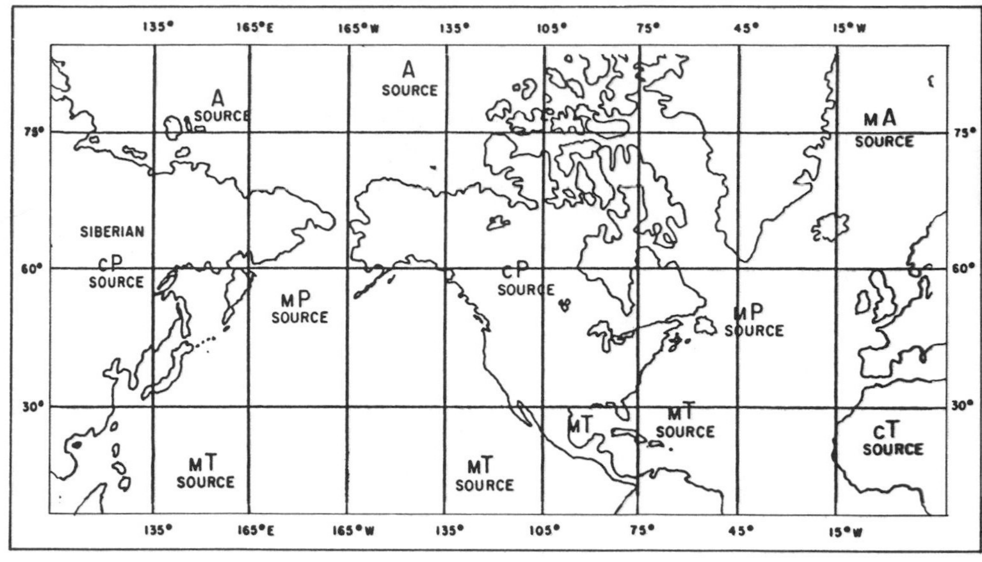

Figure 10-2. Air mass source regions.

air from the arctic that has been cooled by the arctic ice and snow. This air is cold and humid. As it moves southward it is heated from below. As the low level air warms, it expands, becoming lighter than the cooler air above it. It now rises, producing instability up to altitudes of 9,000 to 10,000 feet.

Warm summer air from Siberia or North America invading the North Pacific or the North Atlantic north of about 50°N may be modified by the influence of the sea to produce a mP air mass. In this case, the sea cools this initially warm air from below making it stable. Low stratus clouds and fog are likely to result.

CONTINENTAL POLAR (cP)

These air masses form over the northern Eurasian and North American continents. In winter, since sunshine is sparse and snow is an excellent reflector, air that invades these areas cools rapidly at ground level. This leaves relatively warmer air above it. The resulting inversion layer continues up to between 3,000 and 5,000 feet. In consequence these air masses are highly stable. This stability produces predominantly layer, or stratus, cloud types (Fig. 10-3).

In summer, cP air masses frequently form from the merging of *arctic (a)* air with *maritime polar (mP)* air which then moves over the northern continental areas. After a long passage over continental surfaces this air emerges as *continental polar (cP)* having taken on the characteristics of the surface at the time. This may be anywhere from

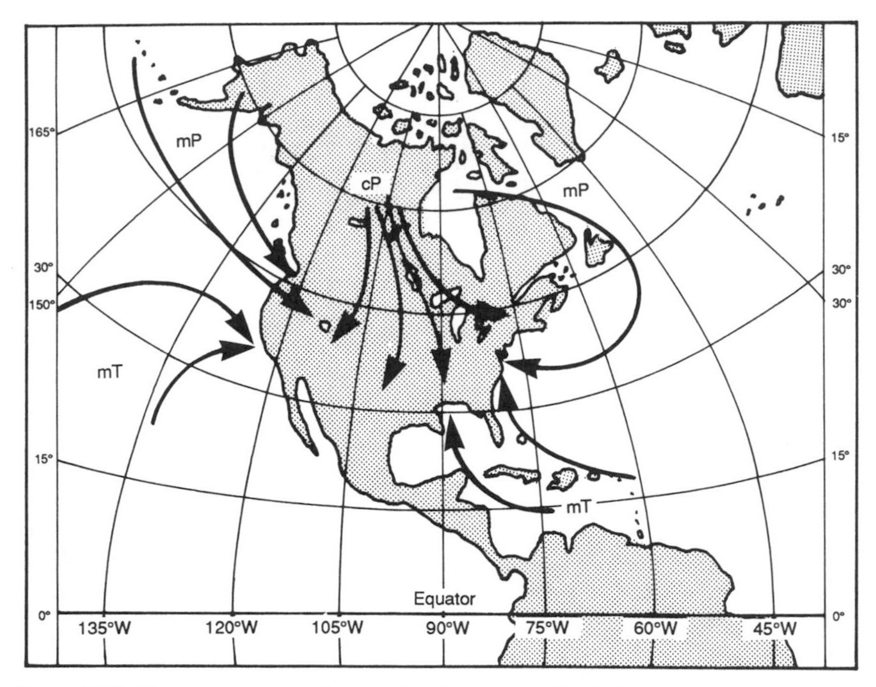

Figure 10-3. Air mass source regions and paths, winter, North America.

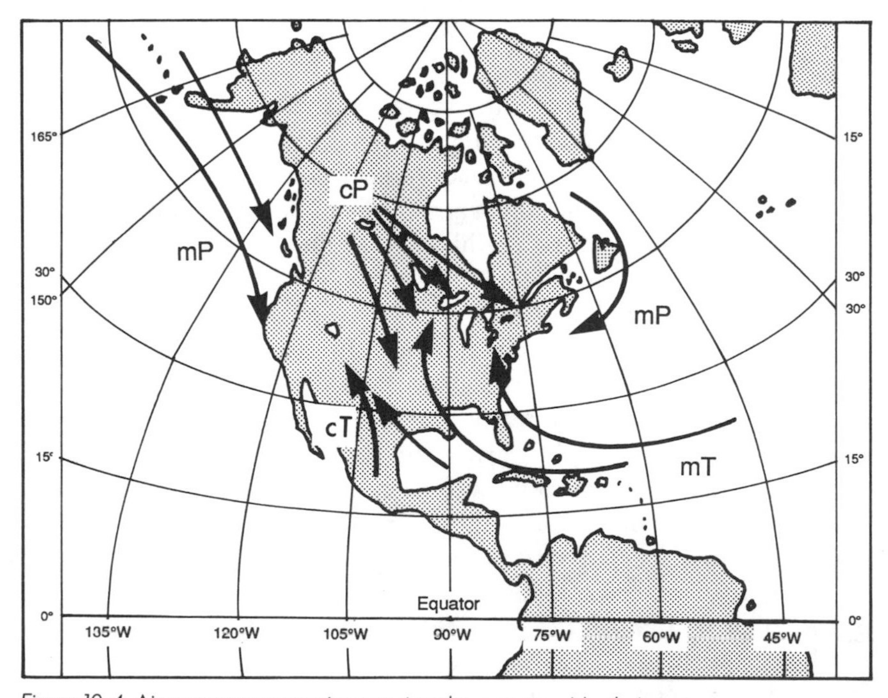

Figure 10-4. Air mass source regions and paths, summer, North America.

cold and stable to warm and unstable. The characteristics vary with latitude and also seasonally from month to month. However, in general summer, cP air masses are moderately cool, have rather low humidity, are unstable at low altitudes, and have cloud bases mostly above 2,500 feet. (Fig. 10-4)

MARITIME TROPICAL (mT)

In winter these masses form between latitudes 20°N and 40°N in the North Atlantic, and between 12°N and 40°N in the North Pacific. They form whenever a "high" or anticyclone develops in these oceans, which is a frequent occurrence.

Through prolonged contact with tropical ocean waters, these masses become warm, humid, and unstable. As they move north and east from their source regions, they follow long paths over cooler waters. This results in cooling from below. The low layer of dense cool air produces stability. Cool air is denser and heavier than warm, remember?

By the time these masses reach coastal waters, they are commonly characterized by the temperature inversions typical of the U.S. West Coast.

In summer, the source regions expand somewhat in both oceans. In the North Atlantic, the area ranges from about latitude 12°N up to about 46°N. In the North Pacific, the range is about 15°N to 45°N. Summer subtropical anticyclones are likely to be larger and more persistent than their winter cousins. Also sea temperatures are both higher and more even.

Along the eastern side of summer anticyclones, the northerly winds tend to be rather stronger than during the winter. California is on the eastern side of the Pacific high. This produces a summer upwelling of cold water along the coasts of Southern California (Fig. 10-5). There the cooling of the coastal air from below results in persistent low stratus clouds, a low-level temperature inversion, and the well-known Los Angeles smog.

By contrast, on the western side of summer anticyclones the water is warmest heating the air above it from below. Unstable air, cumulus clouds, showers, and thunderstorms are thus typical of summer on the U.S. East Coast which lies to the west of the normal Atlantic anticyclone, sometimes called the "Bermuda high."

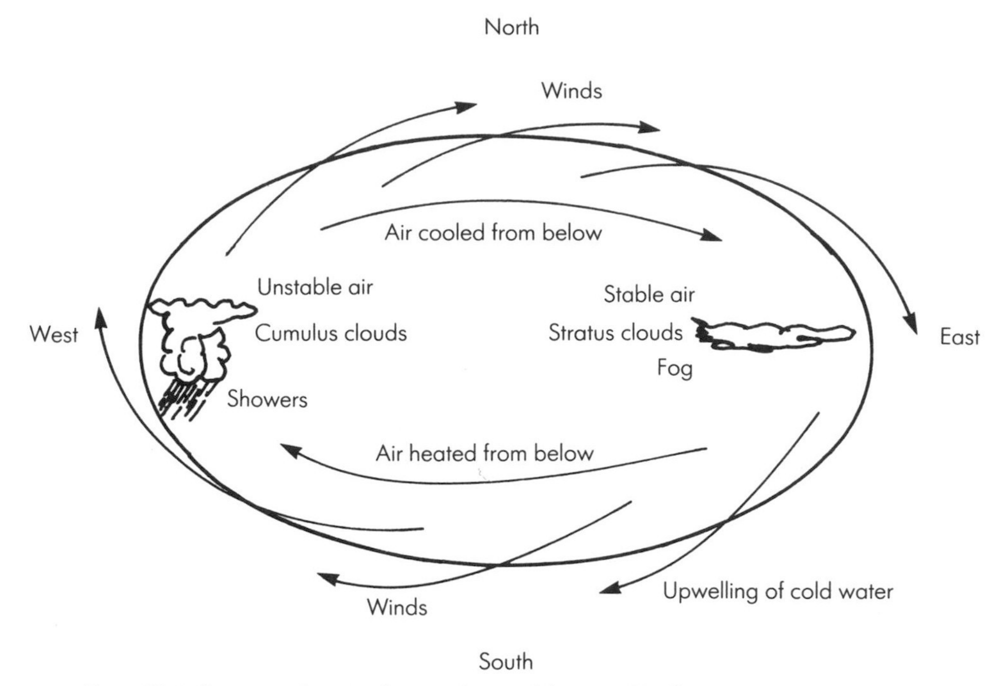

Figure 10-5. Summer subtropical anticyclone—Atlantic or Pacific oceans.

WEATHER FRONTS

As we have seen, a distinctive air mass forms when a large volume of air stalls for a time over one of the several source regions. Large scale temperature and pressure conditions cause these stalls. Eventually, however, the basic planetary air circulation system (see Fig. 8-16) eventually takes over and the entire air mass moves off embedded in the prevailing winds.

Since each air mass has its own distinctive characteristics, it will differ sharply from the air around it as it moves away from its source region. These differences produce clearly defined fronts marking the limits of the mass. Along these fronts the clashing of differences in temperature, pressure, and humidity between huge adjoining volumes of air produce most of the unpleasant weather we experience.

Exactly what kind of unpleasant weather we will encounter when a front passes our vicinity depends on what kinds of air masses are meeting along that front, and how strongly these masses contrast with each other. Although conditions accompanying frontal passages differ considerably, they have some major characteristics in common.

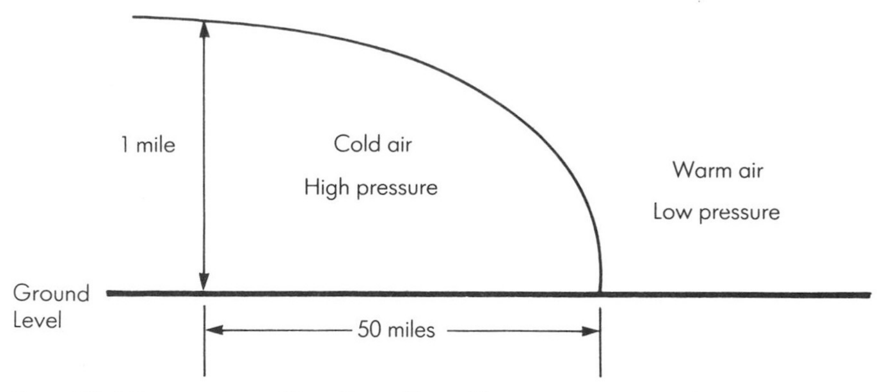

Figure 10-6. Cross section of front formed by cold mass moving toward a warm one.

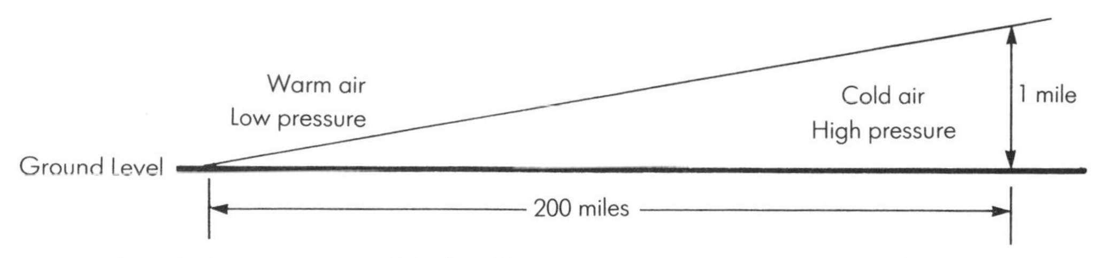

Figure 10-7. Cross section of the front formed by a warm mass moving toward a cold one.

In the atmosphere, frontal surfaces mark abrupt changes in temperature between the air masses on either side of the front, consequently, one mass will always be colder than the other. Cold air is heavier and denser than warm air and thus will show higher pressure readings on the barometer. As a rule the line of separation between any two fluids of different density tends to become horizontal (oil floats on water, right?). Consequently, if the earth were not rotating, cold air (being denser) should simply form in a layer below warm air. However, because of the earth's rotation the line of separation between air masses of different temperatures becomes sloped.

The angle of the slope varies considerably. A cold mass moving toward a warm one (Fig. 10-6) forms a *cold front*. This takes on a fairly steep slope averaging about 1:50. It will reach a vertical height of 1 mile for a run of 50 miles horizontally. A warm mass moving toward a cold one causes a *warm front*. This one has a far more gradual slope (Fig. 10-7). It is likely to be in the range of 1:200. It will reach an altitude of only 1 mile after running 200 miles horizontally.

MAPPING OF WEATHER FRONTS

One of your regular sources of weather information should be the simplified synoptic weather maps as published in daily newspapers (see Fig. 2-14) and shown on TV. On such maps conventional symbols are used to indicate the locations of fronts. A cold front is shown (Fig. 10-8a) by a series of pointed barbs on one side of the line indicating the position of the front. The barbs are on the forward side of that line showing its direction of movement. In the case of a warm front the pointed barbs are replaced by semicircles (Fig. 10-8b) also placed on the forward side of the front.

From time to time cold or warm fronts moving at different rates of speed overtake one another. When this happens, the result becomes an *occluded front*.

Figure 10-8

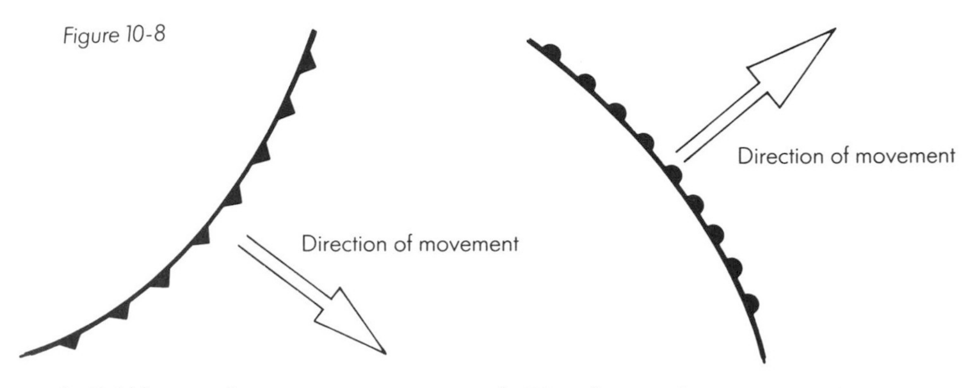

b. Cold front as shown on synoptic weather map.

b. Warm front as shown on synoptic weather map.

Depending on the relative temperatures of the air masses involved, one of two types of occlusions is formed. A *cold occlusion* (Fig. 10-9) occurs when a mass of cold air overtakes a mass of warm air that is already riding up over a third mass of cooler air. In this case there are really two fronts in existence: one is at ground level between the mass of advancing cold air and the receding cool air. The second is 20 to 50 miles behind the first, and at a considerable altitude above the ground. It marks the line between the cold and the warm air masses.

The other type of occlusion is a *warm occlusion*. (Fig. 10-10). This time a cool air mass overtakes two others, one warmer, the other colder. The result is that the cool rides up over the cold mass, which is denser and heavier, but slides under the warm mass which is lighter.

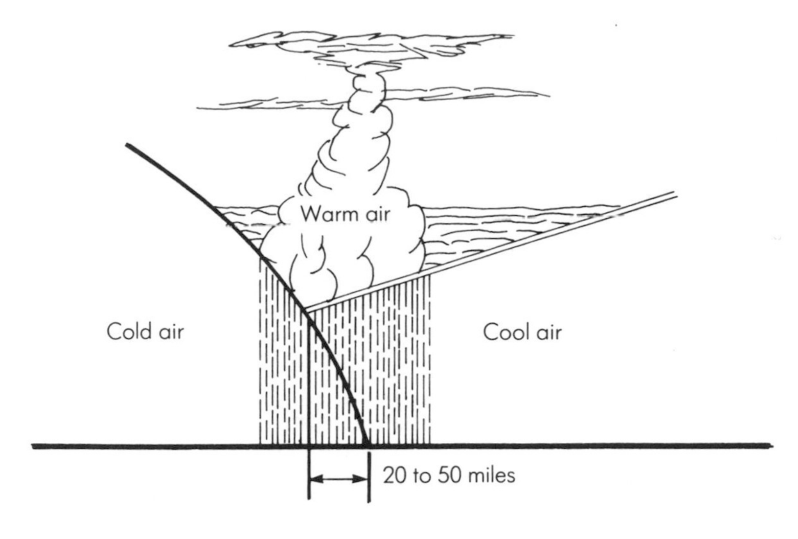

Figure 10-9. Cold occlusion.

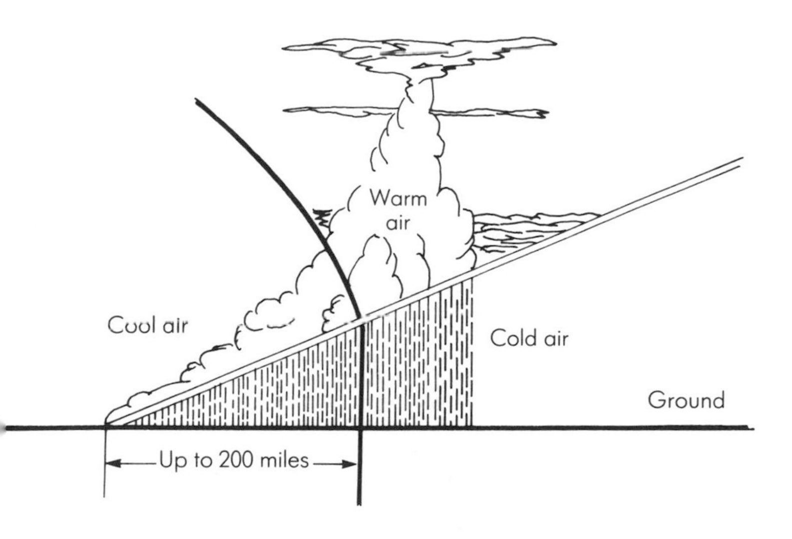

Figure 10-10. Warm occlusion.

Figure 10-11. Map symbol for occluded front.

Again there are actually two fronts. One is at ground level between the cool and the cold air masses while the second is at altitude between the cool and the warm masses. This time the horizontal separation may increase up to as much as 200 miles. The map symbol for either type of occlusion is the same (Fig. 10-11) consisting of alternate barbs and semicircles along the line of the ground level front.

After an air mass has formed over a source region, and then moved on, conditions may arise causing it to stall again. When this occurs, obviously the front defining its edge stalls as well. This results in a *stationary front* which appears on the map as a line with barbs on one side and semicircles on the other (Fig. 10-12).

As we shall see in Chapter 11, there tends to be a consistency over time in the movement of air masses, and their associated fronts. By watching the direction and speed of a front for several days, it becomes possible to predict it's future location with considerable accuracy.

FRONTAL WIND SHIFTS

A front is a line that marks not only a temperature differential but also a wind shift line (Fig. 10-13). Along this line the wind will always veer, meaning it will move clockwise. If the wind is from the southwest in the warm air mass, when the cold front passes it will move to the northwest.

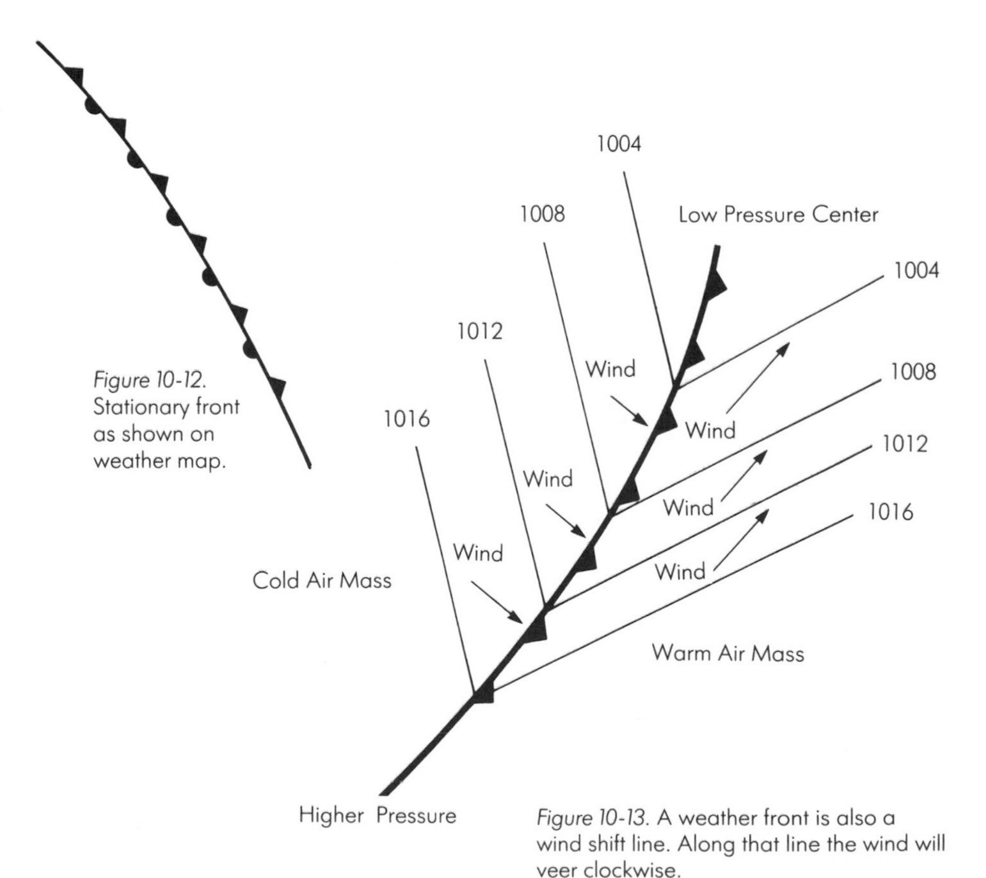

Figure 10-12. Stationary front as shown on weather map.

Figure 10-13. A weather front is also a wind shift line. Along that line the wind will veer clockwise.

FRONTAL WEATHER CONDITIONS

All fronts produce foul weather. How foul depends on how great the differences are between the air masses that meet along a particular front. Strong contrasts produce extremely foul weather, mild contrasts result in only moderately wretched conditions.

Each different kind of front produces its own characteristic weather sequence. These differ in intensity and duration depending on the type of front as well as the strength of air mass differences involved. In general, cold fronts produce rather blustery, turbulent, and sometimes quite violent conditions, but they pass fairly quickly. As we saw in Figure 10-6, they typically extend only about 50 miles horizontally. A fast moving cold front can pass in as little as a couple hours.

Warm fronts often produce weather that is merely gloomy, but they are much slower in passing since they extend horizontally so much further (Fig. 10-7) than do cold fronts. Let us look now in some detail at the sequence of conditions that accompany the passage of each of the different types of fronts.

FAST MOVING COLD FRONTS

Along a cold front, cold air is moving under warm air and driving it sharply upward. This results in a narrow band of comparatively violent weather. The fast rising warm air produces towering cumulus clouds, and heavy rain. Winds are strong and gusty causing choppy seas. Cold fronts often move very rapidly. For a fast-moving cold front a speed of 25 to 30 knots is not uncommon.

A particularly dangerous and unpleasant phenomenon called a *squall line* sometimes precedes a fast-moving cold front. It may appear anywhere from 50 to 300 miles ahead of the front and move parallel to it at about the same speed. When approaching, a squall line appears from a distance as a solid, seething wall of very low black clouds. When squall lines are expected, the mariner will be given warning via Weather Service VHF broadcasts.

With fast-moving cold fronts both the width of the band of bad weather, and its intensity vary depending on the stability of the two air masses that are meeting along the front. The worst possible combination occurs when both air masses are unstable (Fig. 10-14b). In this case towering cumulus and cumulonimbus clouds, rain, gusty winds, and sloppy seas not only accompany the passage of the front itself but they also both precede and follow it for distances of 150 to 200 nautical miles. This means that very actively nasty weather may persist for as much as 12 to 18 hours.

When only one of the meeting air masses is unstable, the period of poor weather is shortened. If an unstable warm mass is overtaken by stable cold air (Fig. 10-14c), squally weather will precede frontal passage by up to a couple hundred miles. However, when the quite blustery front has passed, cool, clear weather quickly follows.

When the unstable mass is the cold one rather than the warm, fine

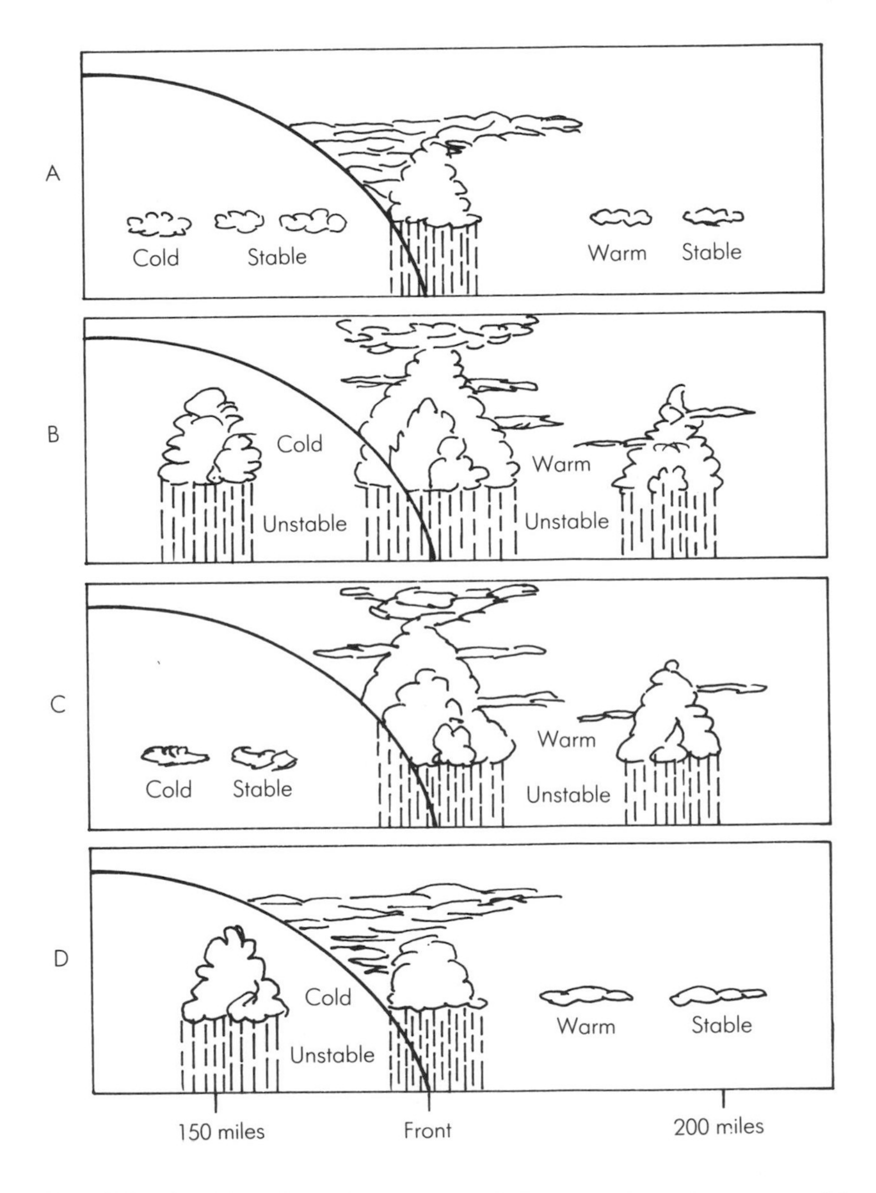

Figure 10-14. Conditions along a fast-moving cold front vary depending on the stability of the two air masses.

weather will hold almost until the front arrives (Fig. 10-14d). The frontal passage is not as turbulent as when the warm air is unstable, but after it passes squalls and rain will persist as the unstable cold air moves in.

Ideally, if you've got to endure the passage of a fast-moving cold front you'd prefer to have both air masses stable (Fig. 10-14a). The squall band at the front is not as wide or as intense as when one or both of the masses are unstable, and the whole thing is over far more quickly.

SLOW MOVING COLD FRONTS

The slope of a slow-moving cold front is considerably more gradual than that of a fast mover (Fig. 10-15). This produces a much wider belt of frontal weather, but less violent than the fast fronts. The typical cold front cumulus clouds take the form of stratocumulus and altocumulus, and appear 100 or more miles in advance of the slow front.

Slow cold fronts move at only about 15 knots. This combined with the more gradual slope of the fronts means that they take considerably more time to pass. Wind and sea will not be as boisterous as with a fast-moving front, but you'll be longer laboring through them.

Basically, when a cold front is due to pass, particularly a vigorous one containing unstable air, it is well for small boats to stay in port. At the very best it will be unpleasant at sea, and at worst it may be

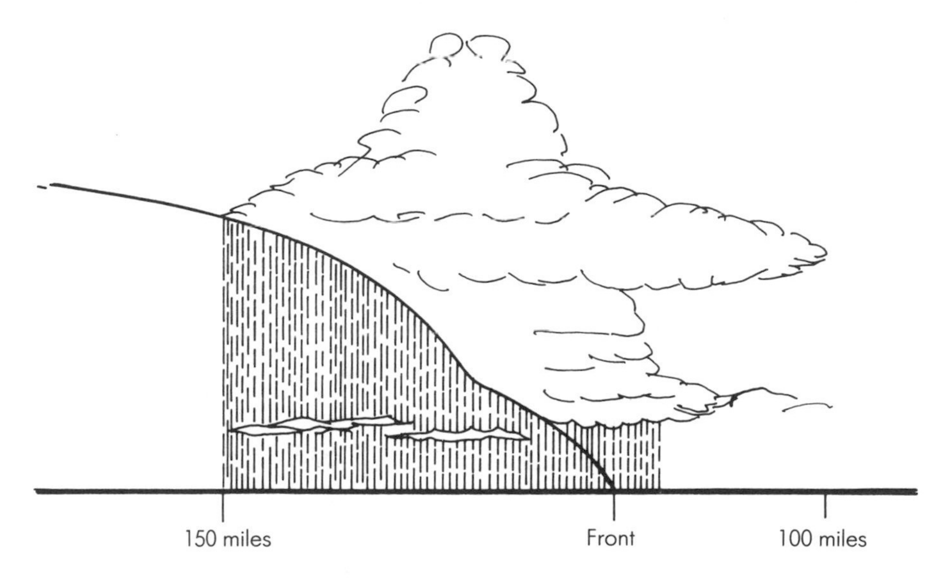

Figure 10-15. Slow-moving cold front.

dangerous. The typical cloud, wind, and weather sequence accompanying the passage of a cold front is summarized in Chapter 11, Table 11-1.

WARM FRONTS

Along a warm front, warm air is replacing a receeding cold mass. The warm air, being lighter, slides up over the heavier cold. (Fig. 10-16) The frontal slope is very gradual, and produces a huge cloud system. Cirrus clouds, the first indicators of a coming warm front may appear as much as 1,000 miles in advance of the front.

After the appearance of the cirrus advance guard, the typical cloud sequence consists of ever thickening and lowering stratus type clouds. This is the immediate visible difference between approaching warm and cold fronts. Cold fronts bring mostly cumulus type clouds, warm fronts produce mostly stratus types.

Warm fronts and their associated weather changes move very slowly. Speeds of 10 to 15 knots are common. With an advance cloud band 600 to 1,000 miles deep and moving very slowly, the first cirrus clouds of an oncoming warm front can often be seen two days in advance.

Following the cirrus come cirrostratus clouds. These, in turn, are followed by lower, thicker altostratus. The front may be as much as 500 miles away at this point. Finally low stratus and nimbostratus move in, and the rain or drizzle starts. When this happens the surface front could still be 300 miles away. Fog is common in the receding wedge of cold air ahead of a warm front.

While the front is moving in, the wind will be east to southeast. When it shifts to southwest or west the front has arrived. At that time the temperature will rise, and the sky will usually clear.

A warm front does not cause the gusty winds and lumpy seas that come with a cold front. While often quite strong, the winds tend to be steady. Conditions on the water are certainly not as dangerous during a warm front passage as during a cold front, but it's going to be wet, dark, dank, and by no means a jolly time. The weather typical during a warm front passage is summarized in Chapter 11, Table 11-2.

COLD OCCLUSION

Occlusion occurs, you will recall, when two cold masses squeeze warm air upward between them. In advance of a cold occlusion, weather conditions are essentially the same as those ahead of a warm front. This is because the cold occlusion is preceded by the typical frontal slope and the normal stratus cloud sequence of a warm front (see Fig. 10-9).

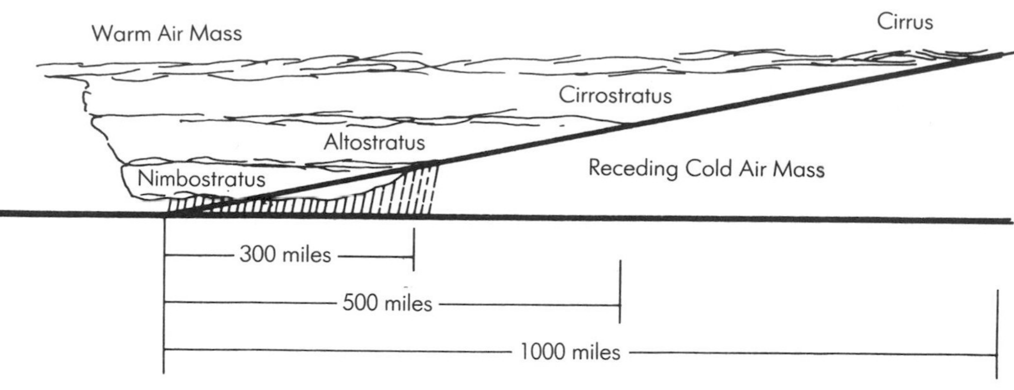

Figure 10-16. Warm front.

When the cold front arrives at ground level, wind shift, pressure, and temperature changes resemble those of a cold front passage. The wind shifts sharply clockwise, pressure rises, and temperature drops. The usual gusty, violent cold front conditions now take over. It is again a good day for the boatman to stay in port.

WARM OCCLUSION

In advance of a warm occlusion conditions are also the same as those in advance of a normal warm front. Cirrus followed by gradually lowering and thickening stratus clouds precede the upper level cool front (see Fig. 10-10). Both the cool and the warm air are riding up over the receding cold mass at the bottom level.

The surface weather sequence is essentially the same as for a warm front passage. Winds shift clockwise, pressure steadily declines, then levels off, temperature rises somewhat. Rain starts well before the upper level front passes, and continues until after the surface front has moved through. The rain band of a warm occlusion is much wider than that of a cold occlusion. In short, again it is not going to be a good day. It will be chilly, and gloomy, but not blustery and turbulent as with a cold occlusion.

11

TEMPERATE ZONE CYCLONIC STORMS

The extratropical cyclones that occur in the temperate zones (between 30° N and S, and 60° N and S) are by a considerable margin the largest storms on earth. They are by no means the most violent, but they sprawl out over gigantic geographical areas. A single storm often affects thousands of square miles of the earth's surface. Within much of that affected area wind and sea conditions are often such that small boats had best remain in port, and in the rest of the area conditions are likely to be poor at best.

These storms are most intense during the winter. Also they tend to form in strings or "families" of several storms in a row. A family of cyclones may bring a protracted period of bad weather that can last for weeks as a sequence storms passes over.

POLAR FRONT THEORY

The primary atmospheric wind and pressure belts on our planet (see Fig. 8-16) show the polar easterlies blowing into a belt of low pressure at about 60° N where they are met by the prevailing westerlies which cover a band from there down to the horse latitudes at about 30°.

North of the dividing line is cold polar air while to the south is warm air that has crossed the temperate zone from a source region in the tropics. That dividing line is actually a front, called the *polar front* (see Chapter 10) between two gigantic contrasting air masses. At times the easterlies and westerlies blowing parallel flow smoothly

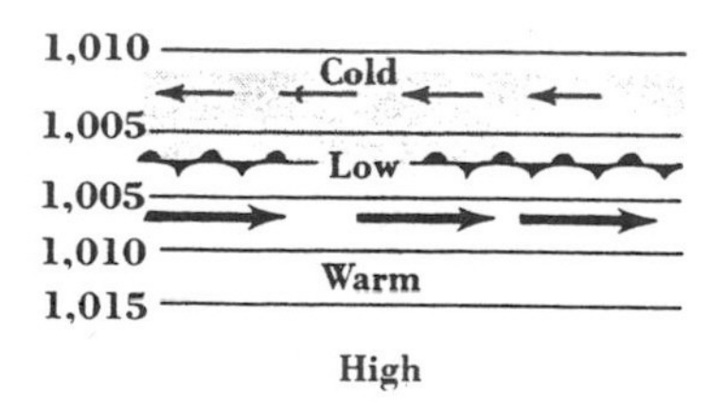

Figure 11-1. Calm condition along the polar front. From: Petterssen, *Introduction to Meteorology*.

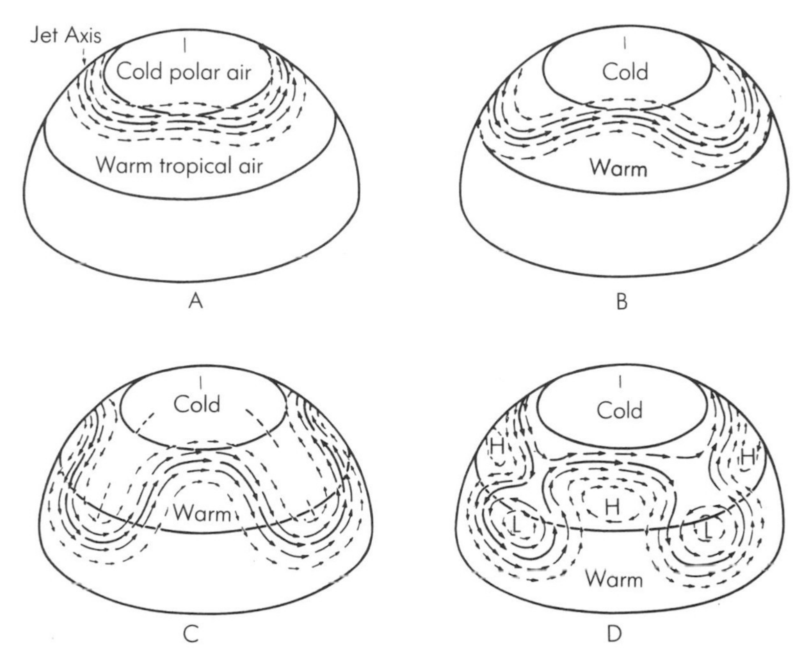

Figure 11-2. a. Polar front becomes unstable and starts to undulate. b. Rossby waves begin to form. c. Strongly developed Rossby waves. d. Waves break up, warm and cold high and low pressure cells form and dissolve. From: Donn, *Meteorology*, from G. H. Trewartha.

past each other and no weather disturbances develop (Fig. 11-1). Then for a combination of reasons this smooth flow becomes unstable (Fig. 11-2a) gradually building up a series of upper level waves called *Rossby waves* (Fig. 11-2b), named after C.G. Rossby, the meteorologist who originally explained them.

These waves consist of tongues of cold polar air that have intruded at upper altitudes into the warmer air mass of the temperate latitudes (Fig. 11-2c). They move eastward embedded in the winds of the prevailing westerly belt they have invaded producing a series of alternating warm and cold fronts. It is from these alternating warm and cold fronts that cyclonic storms are formed. As the storms develop, the cold fronts move faster than the warm ones and overtake them. The circulation now breaks up into alternating cold high pressure, and warm low pressure cells (Fig. 11-2d). These finally mix, fill, merge, and dissolve after which the polar front becomes reestablished. The life cycle of Rossby waves varies from a few days up to a little over a week.

LIFE CYCLE OF A SINGLE CYCLONE

A few extratropical cyclones originate in specific geographic locales such as the fixed thermal low pressure area in our southwestern deserts. However, most cyclones start as instability waves along the polar front. So to trace the life history of a typical cyclone, let us start with an undisturbed section of the polar front (Fig 11-1).

There are marked temperature and density differences between the polar air to the north, and the tropical air to the south. Whenever fluids (we are considering air, for the moment, as a fluid) of different characteristics are in contact, the surface between them is potentially unstable. Along the polar front the polar air, being heavier and denser, pushes under the lighter tropical air. Thus even when the polar front is a relatively straight line horizontally it is sloped vertically. As the cold polar air spreads out under the warmer air it becomes increasingly unstable.

Any one of various factors can start a minor disturbance along this line: irregular topogography such as a mountain range, thermally induced convection, or anything else that produces surface or upper air irregularities in air flow. This minor disturbance now produces a small wave (Fig. 11-3) on the already unstable front. Once started it grows rapidly, and moves eastward along the polar front.

A center of low pressure forms at the apex of the wave which becomes more intense as the wave grows. The wind flow also changes with the formation of the low pressure center (Fig. 11-4). Around the low pressure center circles *(isobars)* of gradually increasing pressure radiate, and the winds spiral inward around those circles. An isobar is simply the term for a line on a map along which atmospheric pressure is the same at a particular time.

At this point the inside of the wave is composed of a mass of warm air moving eastward with cold polar air both ahead of it, and behind it. The forward face of the wave is a warm front because warm air follows it (see Fig. 10-6). The rear face is a cold front since cold air

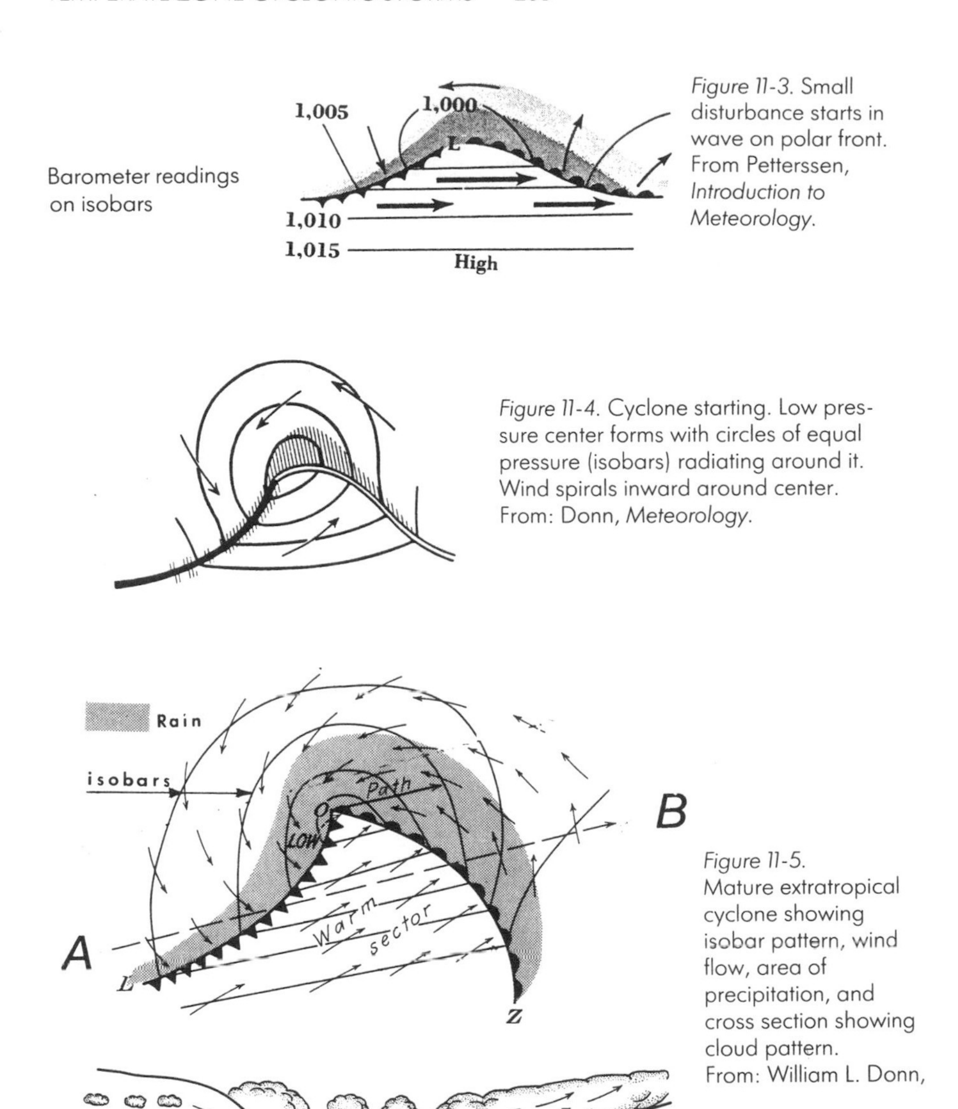

Figure 11-3. Small disturbance starts in wave on polar front. From Petterssen, *Introduction to Meteorology.*

Figure 11-4. Cyclone starting. Low pressure center forms with circles of equal pressure (isobars) radiating around it. Wind spirals inward around center. From: Donn, *Meteorology.*

Figure 11-5. Mature extratropical cyclone showing isobar pattern, wind flow, area of precipitation, and cross section showing cloud pattern. From: William L. Donn,

follows it (see Fig. 10.4). Each front produces its typical cloud sequence as described in Chapter 10.

As the storm moves eastward the cold front moves more rapidly than the warm front. The wave form becomes more pointed and the low pressure center grows and deepens as the storm reaches its mature state (Fig. 11-5). At this stage the typical warm front stratus cloud and rain sequence precedes the arrival of the sector of warm clear air. This followed by cold front cumulus and rain squalls after which the winds shift to the northwest and skies clear.

After a cyclone reaches the mature state, the cold front continues to overtake the warm one. Cold air slides in under the warm sector pushing it upward to form an occlusion (Fig. 11-6). The fronts now gradually dissolve, higher pressure air fills up the low pressure center, the wind circulation weakens, and finally the storm disappears.

THE BUYS-BALLOT LAW OF STORMS

Cyclonic storms travel from west to east. The worst wind and sea conditions develop near their centers. It is helpful if the sailor can locate the direction of a storm center from his position since he can then predict whether trouble is likely to come his way, or pass him by.

A Dutch meterorologist named Buys-Ballot discovered a very simple method to approximate the location of a cyclone center relative to your position. Merely face the wind, and the storm center will be close to 90° to starboard.

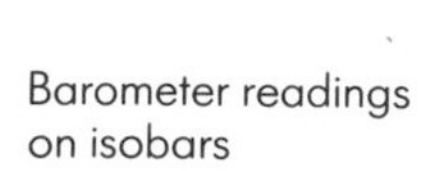

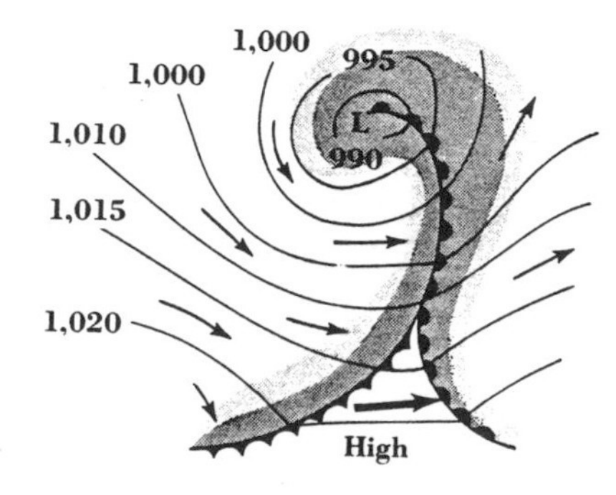

Figure 11-6. Occluded cyclone—most of the warm sector of the mature cyclone has been pushed aloft by the overtaking cold front. From: Petterssen, *Introduction to Meteorology.*

The winds blow counter clockwise and spiral inward around an extratropical cyclone (see Figs. 11-4, 11-5, and 11-6). Due to surface friction, the surface wind spirals inward toward the storm center at a somewhat greater angle than the wind aloft. There will be, therefore, some discrepancy between the direction of the surface wind, and that of the low clouds. Thus, when facing the surface wind the storm center will usually be more than 90° on your right. Probably somewhere between 90° and 120°. So, if instead of the surface wind, you face the direction from which the low clouds are coming the center will now be very close to 90° on your starboard.

EXTRATROPICAL STORM TRACKS

We have seen that extratropical storms start as disturbances along the polar front which divides the cold polar air from warm tropical air. This front is constantly in motion. It moves north and south with the changing seasons. There are calm times when on a map it becomes a fairly straight line. Then it starts to undulate and develop Rossby waves. From these waves cyclonic storms develop. After they spend their fury, they break up and dissolve allowing the front to reform, after which the process is repeated.

Considering both the calm periods and periods of mild to strong Rossby waves, the polar front passes, at one time or another, over just about the entire area of the belt of prevailing westerlies. Since cyclonic storms can form at any place along that front, the entire belt is subject to these storms. However, due to land topography and other factors that influence where disturbances start, some areas are struck more frequently than others (Fig. 11-7).

In general, the northern part of the country, from Washington and Oregon on the west across to New England on the east, gets more frequent and more severe storms than the south. Also extratropical cyclones are more frequent and more severe during winter months than during the summers. The areas of minimum frequency and minimum severity are the southwest (Southern California, Arizona, New Mexico), and Florida. Florida is a likely target for tropical but not extratropical cyclones.

Maritime polar (mP) air masses colliding with warm tropical air blowing out of the Pacific High frequently forms cyclones over the Gulf of Alaska. These then move eastward over Washington and Oregon, and on across the U.S.

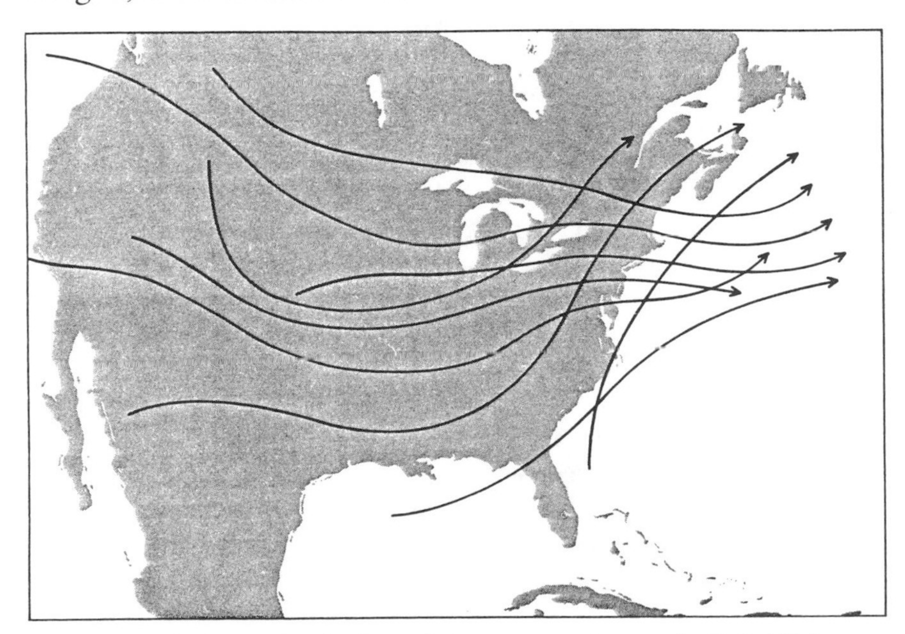

Figure 11-7. Paths frequently taken by cyclonic storms passing over the United States.

Continental polar air (cP) originating in Canada often starts cyclone formation over the southwest and south central states where it collides with maritime tropical air (mT) from the Pacific Ocean, warm air in the thermal low over the southwestern deserts, or maritime tropical air out of the Gulf of Mexico. These storms then move eastward. They are often deflected to the north by the Appalachian Mountain range. In this case they miss the coastal areas of the southeastern states and strike through the northeast before moving out into the Atlantic. Of course, some storms do get through, but the mountains provide a considerable measure of protection to the southeast coast.

CYCLONE FAMILIES

Meteorologists have long known that extratropical cyclones seldom occur singly. Generally an outbreak of polar air into the middle latitudes produces a series of waves along the polar front. Earlier we examined the life cycle of one of those waves.

As a southward intrusion of polar air starts, the cold front takes on a northeast to southwest slant. The first wave develops toward the northeast end of that front. As that wave matures into a full-grown cyclone the polar air to the west of it continues moving south extending the cold front southwestward. Friction between the advancing polar air and the warmer prevailing westerlies produces additional waves. A *cyclone family* of anywhere from two to five members may form in this way.

By the time such a family of cyclones has developed (Fig. 11-8) the leading member is an old occluded cyclone. The second is, or soon will be, partly occluded. Number three is maturing while number four is just starting. As the leading cyclone dissipates new ones may form along the trailing end of the front. For this reason once cyclonic weather moves into an area it is likely to persist with only scattered brief improvements for some time as successive members of the cyclone family pass through.

Just as cyclones do not occur singly, neither do cyclone families. When large Rossby waves form (Fig. 11-9) in the upper atmosphere, commonly in multiples of four, an equal number of cyclone families are likely to form as well. Thus, when storms are raging in the U.S. we often hear that Europe and the Far East are blanketed at the same time, and vice versa.

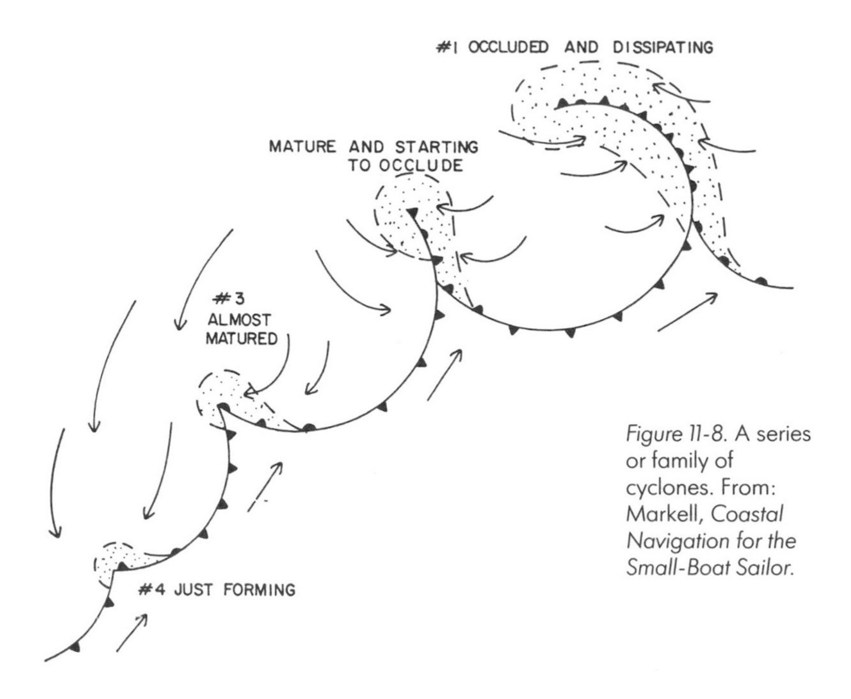

Figure 11-8. A series or family of cyclones. From: Markell, *Coastal Navigation for the Small-Boat Sailor.*

Figure 11-9. Four large Rossby waves aloft have spawned four cyclone families at surface level. After: Palmén

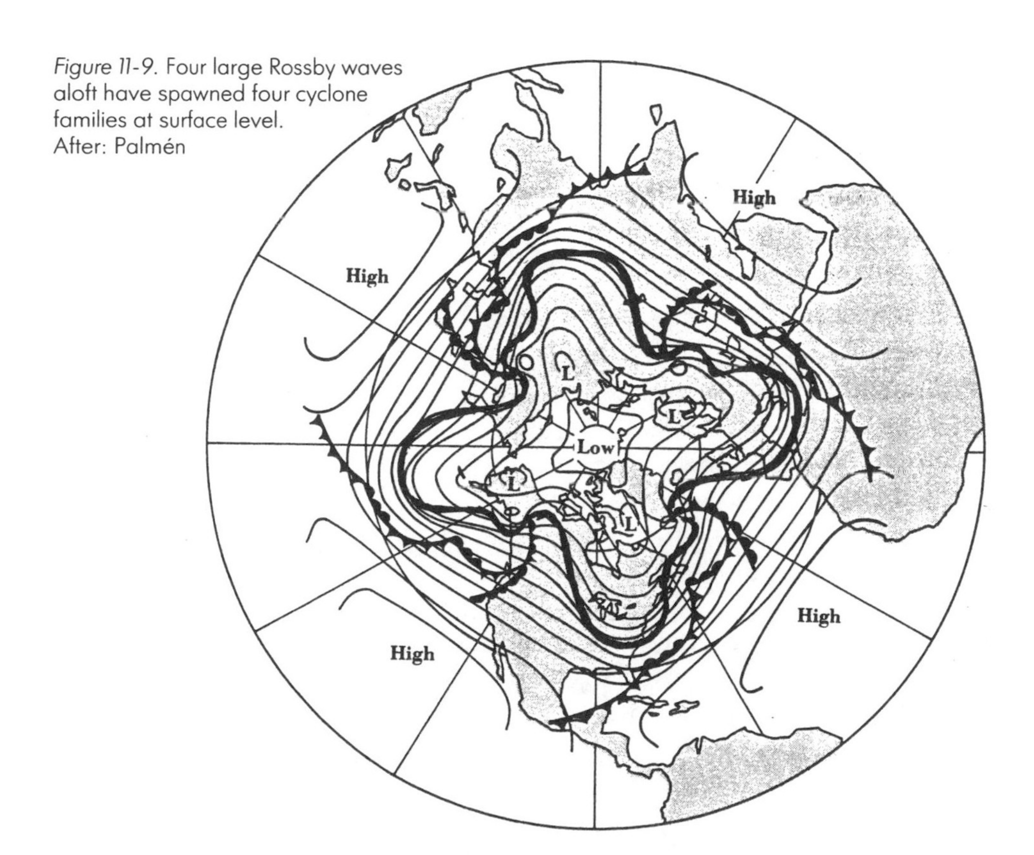

CLOUDS, WINDS, AND SEA CONDITIONS DURING CYCLONE PASSAGE

The conditions you will experience during the passage of a cyclone will vary depending primarily on two major factors;

1. Where you are relative to the low pressure center of the storm—is it passing north or south of you?
2. At what stage of development is the storm when it passes over?

WEATHER SEQUENCE—MATURE CYCLONE—CENTER TO YOUR NORTH (Fig. 11-5)

Typical Warm Front sequence discussed in Chapter 10, clouds are stratus types starting with high cirrus, then thickening and lowering to cirrostratus, altostratus, stratus, nimbostratus and rain. Barometer has been steadily falling. Temperature remains steady. Winds back to the south then on through southeast toward east and increase. Seas building steadily with increasing wind.

WARM FRONT PASSAGE

Barometer becomes steady. Temperature rises. Clouds become scattered small cumulus type. Sky mostly clear. Winds shift to southwest and decrease. Seas decrease as well.

COLD FRONT PASSAGE

Barometer drops then rises abruptly with frontal passage. Temperature abruptly drops. Clouds quickly increase to heavy cumulus and cumulonimbus. Winds shift to northwest becoming gusty and squally. Heavy rain squalls and sometimes thunderstorms. Seas become steep and lumpy in squalls. Cold front, rain squalls, and heavy clouds pass quickly leaving fair weather cumulus clouds and gradually decreasing gusty winds in their wake.

WEATHER SEQUENCE—MATURE CYCLONE—CENTER TO YOUR SOUTH (Fig. 11-5)

In Advance Cirrus clouds are followed by lowering stratus types: cirrostratus, altostratus, stratus, and nimbostratus and rain. Winds move to southeast then back through east to northeast and increasing. Seas increasing. Barometer drops steadily with little change in temperature.

CENTER PASSES TO SOUTH

Rains decrease, become showery, and finally stop. Winds back to north, then northwest decreasing and becoming puffy. Temperature slowly lowering, barometer slowly rising. Seas decreasing. Clouds lifting, thinning, and changing to scattered cumulus types.

WEATHER SEQUENCE—OCCLUDED CYCLONE—OCCLUDED NORTHERN PART (Fig. 11-6)

In an occluded cyclone most of the warm sector (Fig. 11-5) of the mature storm has been squeezed aloft between the cold air in front, and that behind. The occluded portion looks like either Figure 10-9 or 10-10 depending on where the coldest air is.

In either case the cloud and weather sequence in advance of the occlusion is the same as what preceded the mature cyclone warm front. The difference occurs when front arrives.

COLD OCCLUSION (Fig. 10-9)

The rain band is comparatively narrow. Temperature drops when the cold surface front arrives. Barometer rises. Winds become gusty and veer toward the northwest. The preceding steady rains become showery and squally, then decrease. Skies clear leaving scattered cumulus clouds, puffy winds, and lumpy seas that gradually flatten. Essentially the warm front weather sequence changes right over to the cold front sequence without any gap between.

WARM OCCLUSION (Fig. 10-10)

The rain band is wide similar to that of a normal warm front, and the entire cloud and weather sequence is similar as well. After passage of the surface front, the barometer gradually rises, skies clear, temperature rises, winds veer to the northwest decreasing, and seas decrease.

Winds and sea conditions during the passage of an occlusion are not as severe as those generated during the mature cyclone stage. An occluded cyclone has already begun to dissipate.

WEATHER SEQUENCE—OCCLUDED CYCLONE—UNOCCLUDED SOUTHERN PART (Fig. 11-6)

The wind and weather sequence through the part of a cyclone that has not yet become occluded is essentially the same as that through a mature cyclone when the center is to your north. Typical warm front stratus clouds, steady winds, rising seas, falling barometer, rising temperature, and rain precede a now smaller warm sector. This is followed by the cold front with heavy cumulus clouds, squally winds and rain, rising barometer, falling temperature, and finally clearing with puffy northwest winds, and decreasing seas.

WEATHER SEQUENCE—IMMATURE CYCLONE

The types of clouds, wind, weather, and sea conditions produced during the passage of an immature cyclone (Figs. 11-3 and 11-4) are less intense but otherwise the same as those of a mature cyclone. When the low pressure center is to your north, expect two spells of poor weather—one with each front. When the center is to your south expect one protracted bad spell.

WEATHER SEQUENCE—CYCLONE FAMILY

The passage of a cyclone family is certainly going to bring poor weather with scattered breaks for up to a couple weeks. When the first one in the series to arrive is already occluded (see Fig. 11-8) things are going to get a lot worse before they get any better. If, however, the first one to strike is a mature one, then, while you can expect to get hit repeatedly, each hit will be less intense than the last until the series finally peters out.

Following are two convenient reference tables summarizing the conditions that accompany the passages of the cold front and the warm front during a mature cyclone.

TABLE 11-1.
SUMMARY OF CONDITIONS ACCOMPANYING COLD FRONT PASSAGE

	BEFORE	DURING PASSAGE	AFTER
Weather	Usually some rain Sometimes thunder	Heavy rain—maybe thunder and hail	Heavy rain briefly, then fair—at times scattered showers
Clouds	Altocumulus or altostratus and nimbostratus, then heavy cumulonimbus	Cumulonimbus with Fractostratus or scud	Lifts rapidly followed by altostratus or altocumulus with cumulus later
Winds	Increasing and squally	Abrupt clockwise shift—very squally	Gusty
Barometric Pressure	Falling—moderately to rapidly	Rises suddenly	Continues rising slowly
Temperature	Steady—may drop somewhat in prefrontal rain	Drops suddenly	Continues dropping slowly
Visibility	Generally poor	Poor followed by rapid improvement	Generally good except in scattered showers

TABLE 11-2.
SUMMARY OF CONDITIONS ACCOMPANYING WARM FRONT PASSAGE

	BEFORE	DURING PASSAGE	AFTER
Weather	Steady rain	Rain generally stops	At times light rain or drizzle
Clouds	Cirrus followed by cirrostratus, altostratus, nimbostratus, sometimes cumulonimbus	Low nimbostratus and scud	Stratus or stratocumulus—occasionally cumulonimbus
Winds	Gradually increasing	Shift clockwise—sometimes decrease	Steady
Barometric Pressure	Falling steadily	Holds level	Little change
Temperature	Level or slowly rising	Gradual steady rise	Slow rise or no change
Visibility	Good except in rain	Poor—mist or fog	Fair to poor—mist or fog may continue

12

VIOLENT STORMS

The most awesome weather displays in nature are the heavy rumblings and brilliant flashes of thunderstorms, the screaming fury of tornados, and the shrieking winds, torrential rains, and tumultuous seas whipped up by hurricanes. Compared to the extratropical cyclone (Chapter 11) all of these storms are of vastly greater violence and intensity. The areas they affect are far smaller, but within these smaller areas they rage with stupendous force.

Waterspouts, the marine version of tornados, are an exception. They vary considerably in intensity. One may pack the full violence of a tornado, another may be no stronger than the common dust whirl, or "dust devil," often observed ashore.

Hurricanes move very erratically making accurate prediction of their future movements difficult. However, they are large enough to appear clearly in satellite pictures. This makes it possible to monitor their actual locations and progress quite precisely.

Both tornados and thunderstorms, on the other hand, are individually comparatively small in area, and of brief duration. Thunderstorms often can be seen in satellite observations, and also on radar. Radar detection of thunderstorms is extremely helpful to aircraft pilots and aircraft controllers who are in direct contact with Weather Service. However, these storms are so small, and last such a short time that when one shows up on satellite or radar it is usually too late to warn mariners or the general public as to its probable path. The best the Weather Service can and does do is to issue regional warnings when conditions are favorable for the formation of either tornados or thunderstorms.

THUNDERSTORMS

In spite of the best of precautions sooner or later most sailors finally encounter a violent storm of some type at sea. For the average small-boat operator the most dangerous type is the thunderstorm. It is the most common, and also it is the one he is most likely to encounter without adequate advance warning other than his own onboard observations.

Thunderstorms range in size from single storm cells of perhaps a mile in diameter up to multiple celled storm clusters, or lines of storm cells extending up to fifty miles or more. Meteorologists divide thunderstorms into three classifications: air-mass thunderstorms, line thunderstorms, frontal thunderstorms.

AIR MASS THUNDERSTORMS

Within a fairly uniform air mass (Chapter 10) local "hot spots" can develop due to the differences in heat absorption and reflection between wooded areas, or grasslands, and barren earth (sand or rocky ground). Over such isolated hot spots strong vertical air movement begins the process of *air-mass thunderstorm* formation. This type of storm usually occurs during the mid to later hours on summer afternoons. They then often drift out over the water to disturb a sailor's otherwise peaceful afternoon. Single air-mass thunderstorm cells occasionally develop over an isolated hot spot, but more commonly they appear in clusters. These clusters occur in irregularly scattered locations (Fig. 12-1).

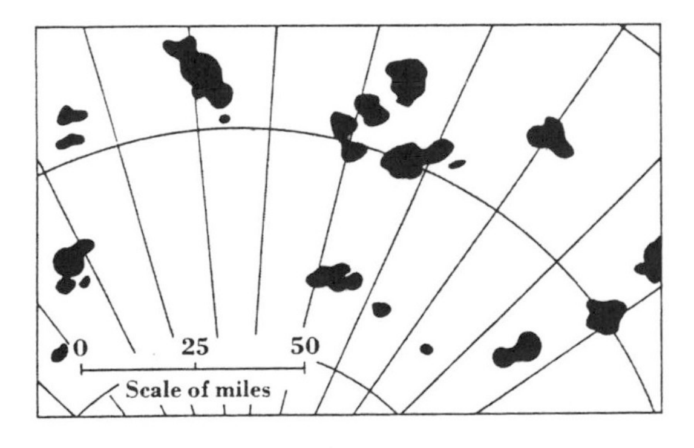

Figure 12-1. Radar view of scattered air-mass thunderstorms. From: Petterssen, *Introduction to Meteorology.*

LINE THUNDERSTORMS

These are multiple cell storms in which the individual cells form in lines or bands following the direction of the low level wind (Fig. 12-2). This type is not often encountered by the saltwater sailor because they most commonly form over land, occurring frequently over the prairie states. While they may develop at any time of the day, they appear most often in the afternoon.

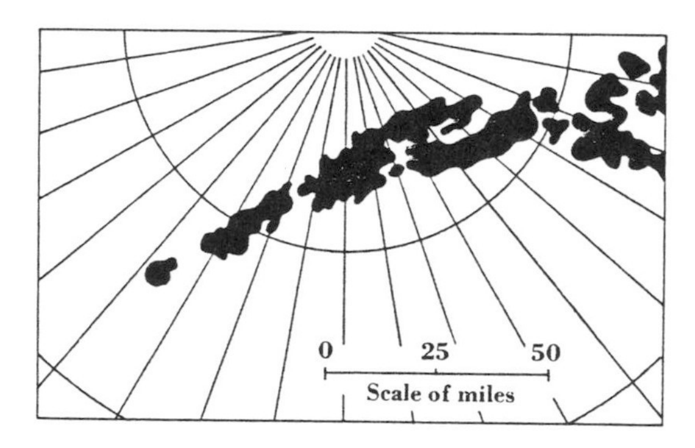

Figure 12-2. Radar view of line thunderstorms. From: Petterssen, *Introduction to Meteorology.*

FRONTAL THUNDERSTORMS

Weather fronts were discussed in detail in Chapter 10. *Frontal thunderstorms* are associated with cold fronts only. When unstable warm air is lifted by an oncoming cold front (Fig. 10-14), thunderstorms may develop in that rising warm air. These storms will be scattered along the front, and will move with it embedded in the general band of frontal clouds. Both other types of thunderstorms prefer afternoons. This type does not. Since they move with the front, these storms may develop at any time of the day or night.

LIFE CYCLE OF A THUNDERCLOUD

The development and dissipation of a thunderhead, seen from a discreet distance, is a very impressive and beautiful spectacle. Figure 12-3 records the growth, maturity, and dissipation of a single thunderstorm cell seen from the dock at a marina in San Diego. This entire sequence occurred in less than one hour. At the time this cell started to develop, another cell to its right had already gone through its entire life cycle and was in an advanced stage of dissipation. The life span of that cell was also less than an hour.

The life spans of these particular individual cells was quite brief. A single cell may, at times, last as long as a couple hours, but thunderstorms are intrinsically storms of short duration.

A single thunderstorm cell starts with a strong updraft in unstable air which develops into a tall cumulus cloud (Figs. 12-3a and b and

Figure 12-3.

A. Thunderstorm developing.

B. Approaching mature stage.

C. Mature thunderstorm.

D. Mature thunderstorm starting to dissipate.

E. Mature thunderstorm dissipating.

F., G., H. Thunderstorm dissipation

G.

H.

Fig. 12-4). The air within the cloud is warmer than the surrounding air causing it to rise rapidly to great altitude. It quickly reaches heights where the temperature has dropped to well below freezing.

Water droplets, snow, and ice crystals have been condensing from the water vapor in the rapidly rising air column. Eventually more water has condensed than the updraft can continue to support so the water then starts to fall down through the cloud. This falling water reverses part of the updraft changing it to a sharp downdraft (Fig. 12-3c and Fig. 12-5). The cloud now consists of an area of rapidly rising air warmer than its surroundings plus a second area of rapidly descending air that is cooler than its surroundings. At surface level the downdraft spreads out ahead of the storm (Fig. 12-6) bringing brief but very intense and dangerous wind gusts. These strong winds plus torrential rain mark the mature stage of a thunderstorm cell.

The torrential rain of the mature stage quickly exhausts the water supply of the storm cloud. The intensity of the rain decreases, and the cloud starts to dissipate (Fig. 12-3e,f,g, and h and Fig. 12-7). By this time upper altitude winds have started to flatten the top of the cloud forming the typical thundercloud anvil top.

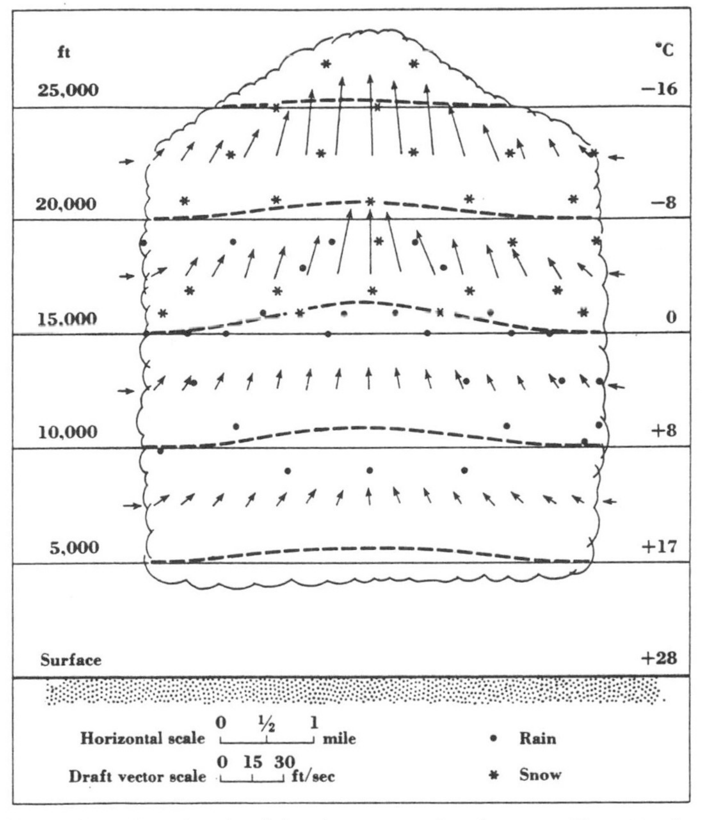

Figure 12-4. Cumulus cloud developing into thunderstorm. The air in the cloud is warmer than its surroundings producing a strong updraft throughout the cloud. See Figures 12-3a and b. From: Beyers, *General Meteorology*.

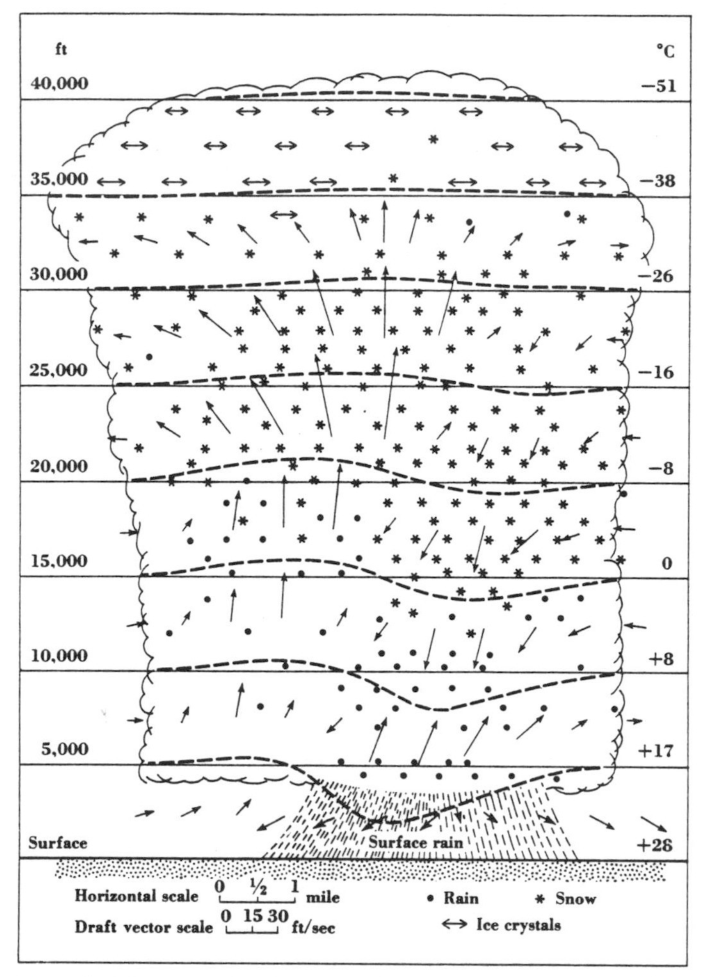

Figure 12-5. A mature thunderstorm cell showing continuing updraft on the left side by side with the downdraft and rain on the right. See Figure 12-3c. From: Beyers, *General Meteorology.*

LIGHTNING AND THUNDER

The brilliant lightning flashes and accompanying thunder result from very strong electrical charges that build up in the thundercloud during its rapid development. These charges are produced by intense friction between the air in the cloud and water vapor that has condensed in the form of water droplets, snow, and ice crystals or hail.

Positive charges cluster toward the upper part of the cloud with negative charges clustering in the lower portion (Fig. 12-8). When the difference in electrical potential becomes large enough, a discharge visible as lightning occurs. The thunderclap that follows results

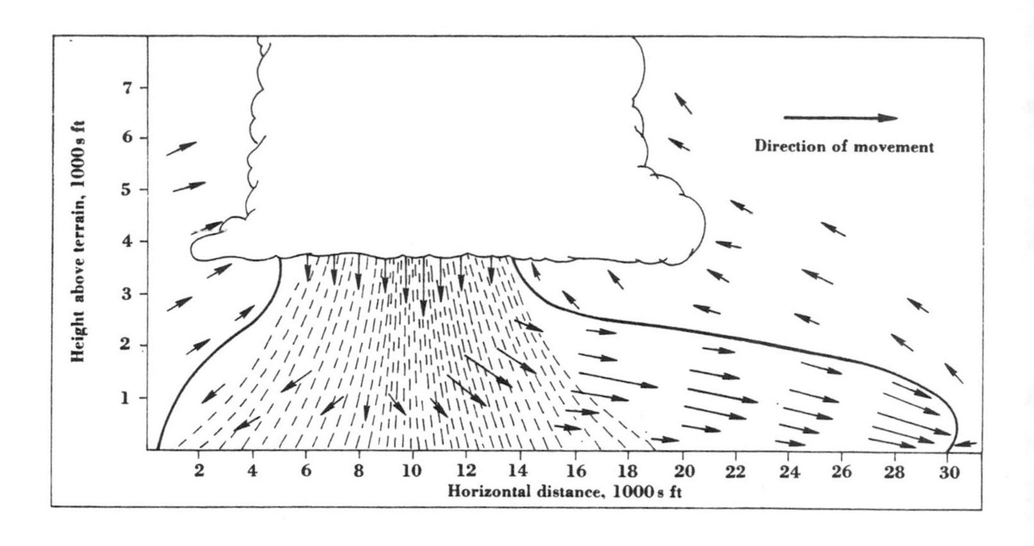

Figure 12-6. ABOVE. Detail of downdraft side of thundercell showing advance cold windgust and rain. From: Beyers, *General Meteorology.*

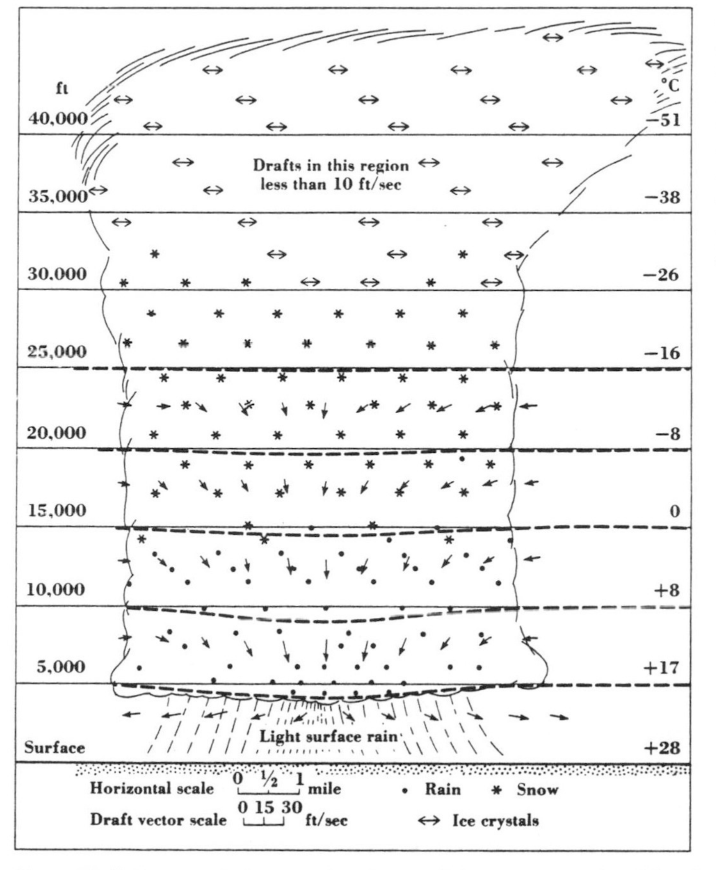

Figure 12-7. Heavy rain has nearly exhausted the water supply of the cloud which is now starting to dissipate. See Figures 12-3e, f, g, and h. From: Beyers, *General Meteorology.*

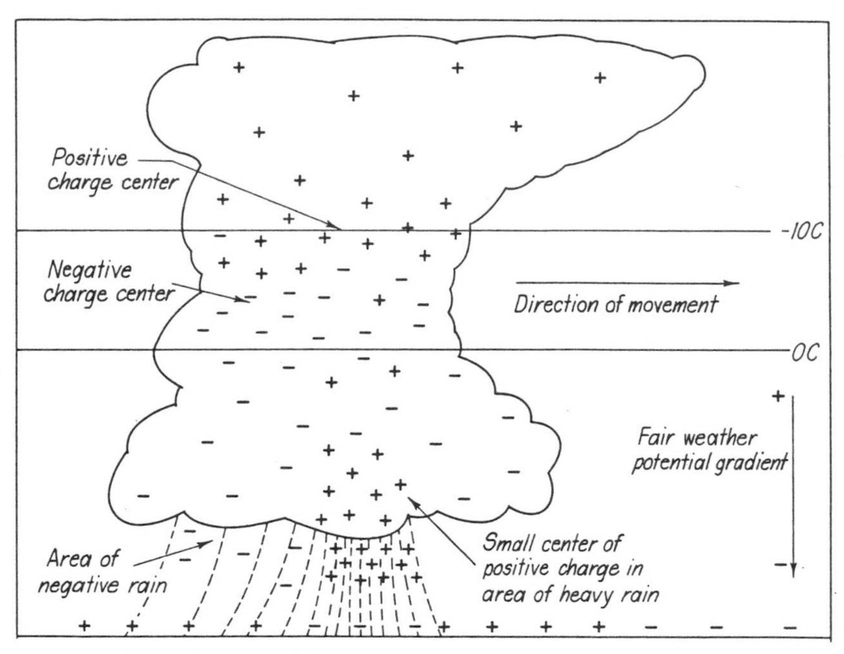

Figure 12-8. Typical pattern of electrical charges in a thundercloud. From: Donn, Meteorology.

from the rapid expansion and contraction of air heated by the passage of the lightning bolt. Sharply defined streaks of lightning are likely to be quite close. *Sheet lightning* is the general illumination of an area of sky by lightning streaks obscured by clouds or down beyond the horizon hence it is likely to be produced by a storm that is relatively far away.

A rough estimate of your distance from a thunderstorm can be made by timing the interval from the time you see a lightning flash, and the moment you hear thunder. Sound in the atmosphere travels a mile in just a bit under 5 seconds so if you count the seconds between seeing lightning and hearing the thunder and divide the total by 5 you have the approximate distance of the storm. Another gauge is that a very sharp, abrupt thunderclap is usually very close. Low, muffled, rumbling thunder is probably far away.

Sometimes at sea in thundery weather, a considerable difference in electrical potential develops between a vessel and the air above it. This may cause bluish streaks to leap from masts or spars, or cause balls of lightning to roll along masts or rigging. This is the famous St. Elmo's Fire that startled and terrified sailors in days of square rigged sailing ships.

THUNDERSTORM CLUSTERS

As mentioned earlier thunderclouds very seldom occur singly. Normally they develop in groups or clusters. To you on the surface, the individual thunderclouds making up a cluster are likely to be indistinguishable. The several clouds in a cluster appear to flow into one another (Fig. 12-9).

On radar they show up as a single large irregular mass. In Figure 12-9 the oldest cloud in the group is on the extreme left. Only a downdraft remains. The newest cell is at the top, completely in the updraft stage. The other three are in various stages of maturity containing both updrafts and downdrafts. Figure 12-10 shows this clearly in crossections A-A' and B-B' through the cluster.

In a cluster such as this new cells tend to form forward of the downdraft side of a mature older cell. Figure 12-6 shows why. The advance cold wind gust ahead of a mature storm cell pushes the warm air in front of it at surface level upward starting the next updraft cycle in motion.

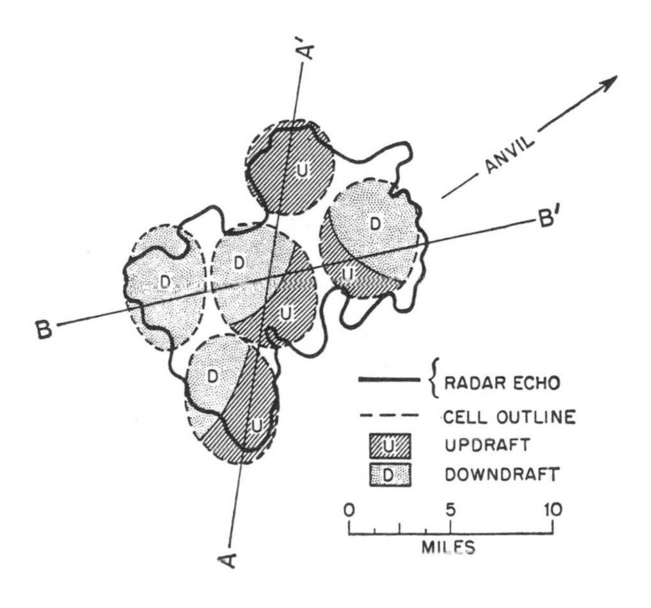

Figure 12-9. Thundercloud cluster of five showing the individual parts and the outline of the groups as it would appear on radar. From: Beyers, *General Meteorology.*

THUNDERSTORM PRECAUTIONS

If you are caught afloat by a thunderstorm the most dangerous aspect is going to be the heavy wind gusts that precede the rain squall. If you are under sail don't be a hardhead, get those sails down until the wind has passed. A few years ago a sailing regatta on Long Island Sound was hit by a rather strong thunderstorm. That advance wind

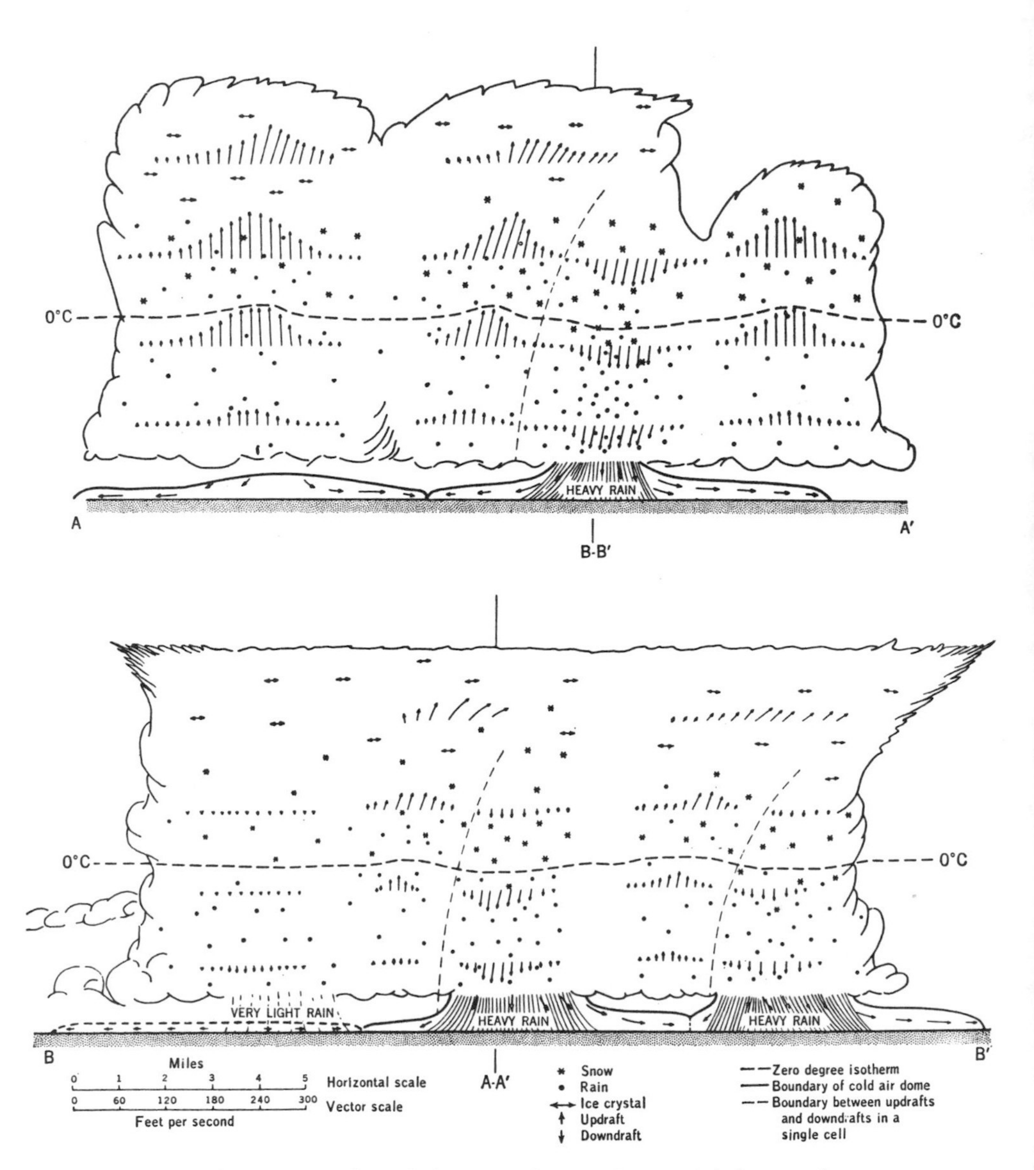

Figure 12-10. Cross section through the storm cluster in Figure 12-9 showing the activity inside the various clouds. From: Beyers, *General Meteorology.*

bowled over better than 100 sailboats in no more than 5 minutes. No regatta patrol is set up to handle a mess like that.

A power vessel is not in as much danger from the wind as one under sail. A power boat, particularly a small one should be careful not to get caught broadside by a sudden wind driven swell and broached.

Although feared by man, lightning is seldom a problem. Much of the lightning we see are bolts discharging back and forth in the clouds. Comparatively few bolts hit the surface of the earth, and a miniscule number of those have ever hit boats. St. Elmo's Fire, mentioned above, may sparkle in the rigging, but it does no damage.

TORNADOS

Of all of the various types of storms that plague us the tornado is far the smallest in geographic area and at the same time by a huge margin the most violent. For destructive fury no other storm even begins to compare. Happily, the path of one of these demons is so narrow that the total amount of damage it can do is considerably less than that done by storms that are far less violent but cover a much larger area. However, everything unfortunate enough to lie within the area a tornado affects will probably be totally destroyed.

The width of a tornado funnel at surface level may very from 20 or 30 yards up to as much as a mile; however, about 300 yards is average. Once they touch down the surface paths of tornados average between 3 and 6 miles long. In rare cases extremely long paths exceeding 100 miles have been observed. The surface path of a tornado is often erratic hopping up to skip over some areas, and descending again to destroy others.

The destruction wreaked by tornados results from a combination of two factors. The wind speeds are phenomenally high and the atmospheric pressures within the funnel are phenomenally low. Since no wind speed indicator (anemometer), or the structure holding it, has ever survived a tornado, accurate measurements have never been made. Theoretical calculations combined with observations of the wind effects indicate that maximum winds reach as high as 500 mph!

Not only is the horizontal wind velocity tremendous, there are immense vertical wind currents as well. The size and weight of objects picked up by tornados indicate vertical air currents up to 200 mph.

The extremely high velocity of the air whirling around to form the funnel results in a sharp drop in atmospheric pressure at its center.

In many instances the destruction of buildings by tornados has been attributed to this very low air pressure. With normal air pressure inside a building the excessively low pressure outside may cause it literally to explode.

Tornados form over land in the warm south sector of cyclonic storms (Chapter 11). The most common time for tornados is in spring and early summer. They may also form in the wake of a hurricane after it has moved over land and is dissipating. A type of cumulus cloud called *mammatus* (Fig. 12-11) is often associated with tornado formation. Be prepared for the worst if you see clouds of this type in your vicinity.

Since tornados form over land they are dangerous mainly to sailors on inland lakes and rivers. Occasionally they move out over the sea to become waterspouts.

Figure 12-11. Mammatus clouds. From this distinctive bulbous cumulus cloud tornados often develop. U.S. Weather Service.

WATERSPOUTS

Waterspouts are very strange phenomena, and are popularly explained by some very curious myths. Actually there are two types. One, as just mentioned, is a tornado that has moved from the land, where it formed, out over the water. This type is a tornado pure and simple with all of the nasty characteristics that go with it. Consequently, it is extremely dangerous to anyone in a small boat—or a big ship for that matter.

The more common oceanic waterspout is considerably different. It is a far less violent convectional phenomenon that can occur almost anywhere at any time—in temperate or tropical waters, and in fair weather or foul.

It too forms from a heavy cumulus type cloud and is caused by a localized convection current. First a narrow dark cone starts to taper down from the base of the cloud. Directly below this, the surface water becomes agitated. Now, a second cone forms funneling upward from the surface. The two finally meet to become a single, continuous, long whirling tube. The spout moves slowly along with its parent cloud for a while until it thins out and separates again into two parts. The upper moves back into the cloud, the lower falls back into the sea.

The violence associated with this type of waterspout varies considerably. It may be no stronger than a common desert "dust devil," or it may be strong enough to destroy small craft. However, in general it is vastly less dangerous than a tornado.

I recall sighting three waterspouts in one afternoon while crossing Virgin Passage from Puerto Rico to St. Thomas. At one point two were visible at the same time. Although one passed quite close—within about half a mile—we were totally unaffected by it.

HURRICANES

Satellite observations coupled with the rapid availability of the weather information gained from them, enables the small-boat operator generally to avoid being caught at sea in a hurricane. However, if he is located on either the East or Gulf Coast he may repeatedly, during a season, have to secure his boat against oncoming hurricanes. Even then the most elaborate of precautions may not save it.

The full fury of a major hurricane defies description. I have been through several while ashore, and I rode one out at sea north of Cuba. While there is no way I can ever forget that experience, and while I am glad I had it, I hasten to emphasize that it is one to be avoided at any cost if at all possible.

In the West Indies these storms were named hurricanes after the word "huracan" used by the Taino indians. The Taino word was adopted by the Spanish who were the first Europeans we know of to experience these storms. Columbus first encountered one in 1494, and another in 1495.

In the East Indies the word is "typhoon." In Australia it is "willy-willy," and in the Phillipine Islands it is "baguio." All of these are different names for what weathermen term a "tropical cyclone."

The tropical cyclone is very different from the temperate zone extratropical cyclone discussed in Chapter 11. It is much smaller in area, seldom more than 300 nautical miles in diameter. The really intense part is considerably smaller still. When fully developed it is usually much more intense than the extratropical cyclone, and most important it is a warm core storm while the extratropical cyclone is a cold core storm.

The U.S. Weather Service divides tropical cyclones into four classifications depending on their intensity. The same storm may change as it develops so as to go through all four classifications at different times in its life cycle. These classifications are:

1. TROPICAL DISTURBANCE—Low pressure area with beginning surface circulation.
2. TROPICAL DEPRESSION—Low pressure area with closed circulation and winds up to 27 knots (31 mph).
3. TROPICAL STORM—Winds are now between 28 and 63 knots (32 to 72 mph).
4. HURRICANE—Wind speed now more than 63 knots (72 mph).

Each of these storms forms within a single mass of tropical maritime (mT) air (Chapter 10) and they are roughly circular in shape. The frontal clashes between contrasting air masses, typical of middle-latitude cyclones, is notably absent.

Atmospheric pressure drops to extremely low levels toward the centers of these storms. As we saw in Chapter 8 "normal sea level pressure" is 1013.2 millibars, or 29.92 inches of mercury. The record low pressure that has been reported in the center of an Atlantic hurricane was 892 millibars, or 26.35 inches. The average is not a great deal higher. The rate of the decrease in atmospheric pressure is not uniform but becomes suddenly much steeper as the storm center approaches (Fig. 12-12).

The tremendously high winds toward the center of a hurricane are directly related to the steepening drop in atmospheric pressure. The most furious winds blow around the very center or *eye* of the hurricane which is an area usually somewhere between 4 and 30 miles in diameter where it is almost flat calm. The eye is a calm area of rapidly subsiding air, alongside of which stands the *convective chimney* that is the basic power center of the storm (Fig. 12-14).

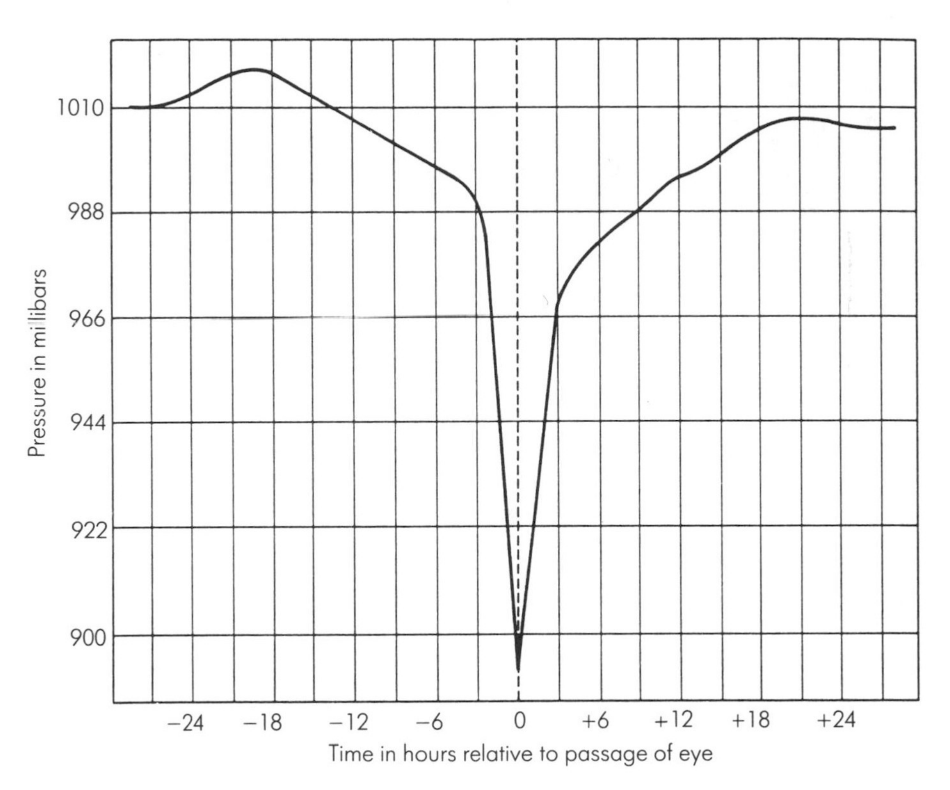

Figure 12-12. Changes in atmospheric pressure during the passage of the eye of a hurricane. From: *Heavy Weather Guide,* U.S. Naval Institute, 1965.

ORIGIN OF HURRICANES

South and east of the Bermuda High, far out in the Atlantic, and deep in the belt of the easterly trade winds a shallow layer of moist air sometimes lies under a deeper layer of subsiding dry, warm air. There is a temperature inversion (Chapter 8) at the level where these layers meet (Fig. 12-13a).

As the winds move to the west (Fig. 12-13b) the inversion weakens, the air layers mix, and convection produces clouds. The typical tropical low pressure trough called an *easterly wave* may form at this stage (Fig. 12-13c).

The easterly wave could be all that develops. However, if convective movement becomes strong enough the easterly wave deepens, the trade winds start to turn on themselves, and a tropical cyclone

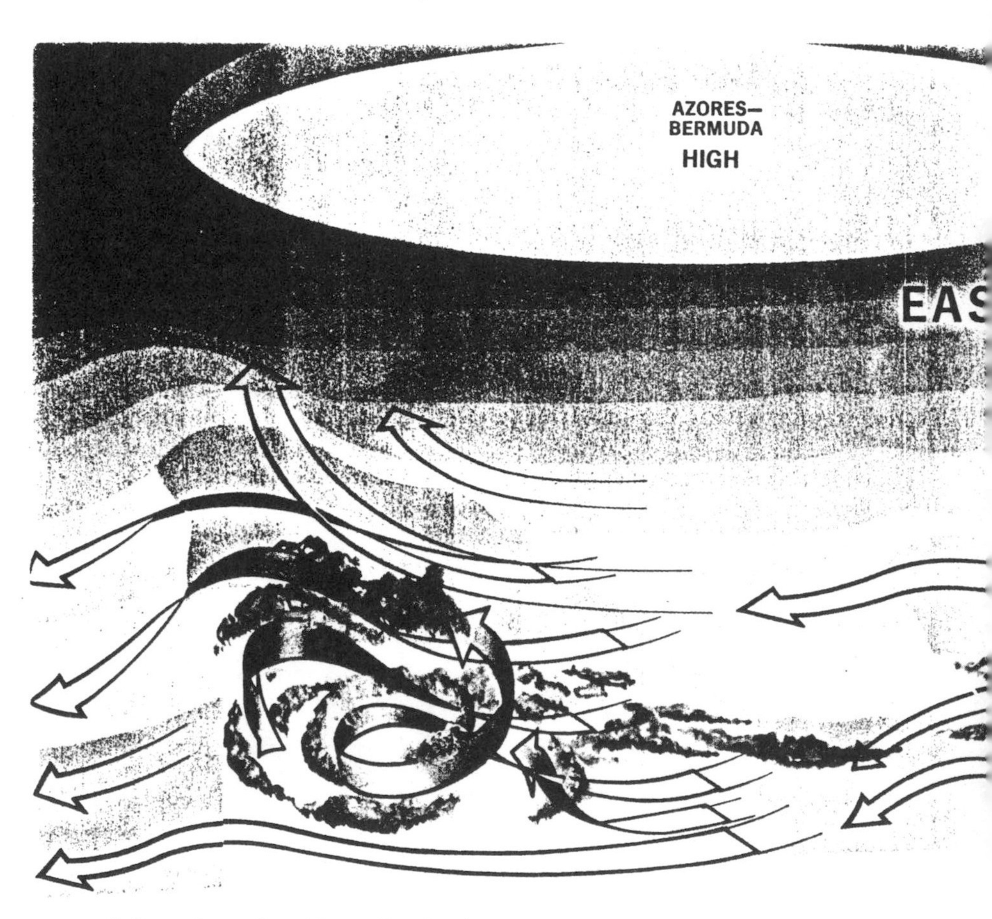

D. Increased convection, and the pumping action of high altitude winds, may cause the easterly wave to deepen further, becoming strong from the surface to 15,000 feet or more. Atmospheric pressure at the surface drops, an area of low pressure becomes an isolated depression — and the trade winds begin to turn in on themselves, forming a weak cyclonic circulation about the center of low pressure. This circulation may intensify, pressure may continue to drop, the winds spiralling around the center of low pressure may accelerate — the disturbance may become a severe tropical cyclone.

C. Intensified convection may be accompanied by a trough of low pressure in the trade wind belt. This poorly developed easterly wave is weak at the surface, stronger near 15,000 feet, weak at high altitude. Thunderstorms may develop behind the wave as the air is raised and shunted northward; ahead of the wave, air is subsiding, weather is fine. A wave of this intensity may travel for thousands of miles without appreciable change.

Figure 12-13. From: *Hurricane,* U.S. Department of Commerce, 1971.

begins (Fig. 12-13d). The tropical disturbance that has now formed may remain no more than that, or it may go through any or all of the rest of the stages up to a full-grown hurricane. Why one easterly wave will travel tremendous distances without developing any further, and another becomes a large destructive hurricane is not yet understood.

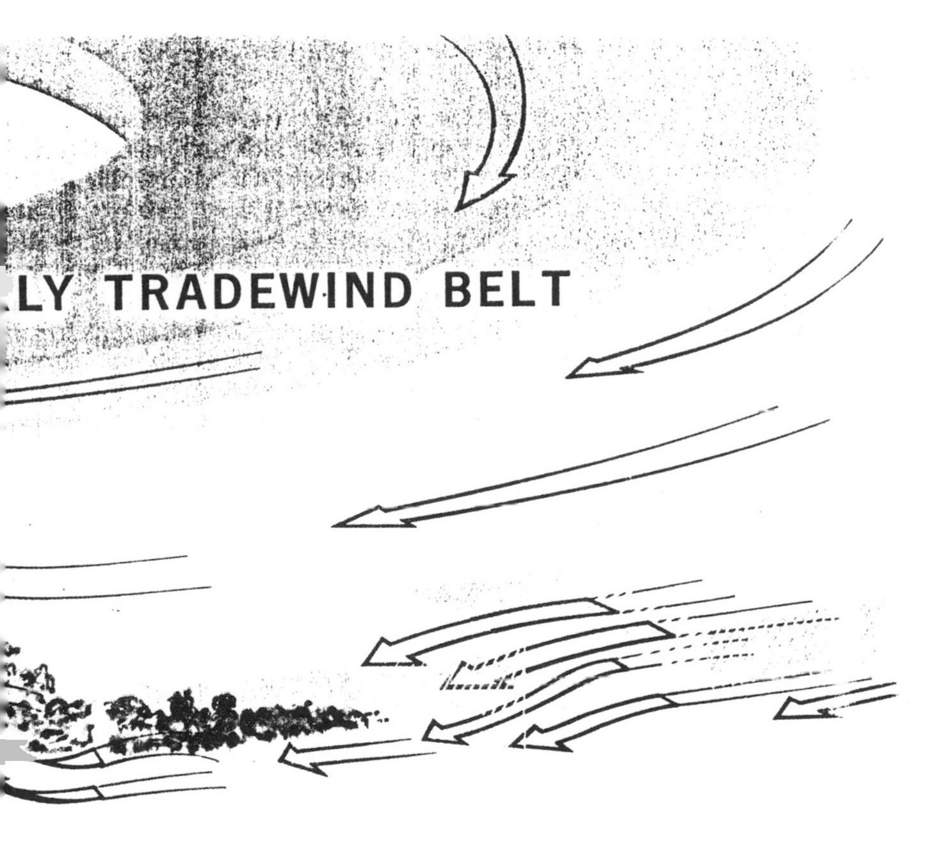

B. As the inversion layer weakens to the west, and as the lower layers become increasingly warm and moist, convection intensifies. Clouds form, and begin to rise to greater altitudes.

A. The characteristic vertical structure of the easterly trade wind belt, in which a relatively shallow layer of moist air is overlain by a deeper layer of subsiding warm dry air. The two layers are separated by a temperature "inversion." Moisture from the sea evaporates into the lower atmosphere but cannot rise above the inversion. However, the moisture charge received by the lower layers contains the latent heat energy that, once released by condensation and convection to high levels, can drive a tropical disturbance.

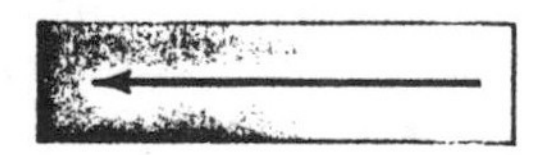

FULLY DEVELOPED HURRICANE

A tropical cyclone is considered a full-scale hurricane when its central winds reach or exceed 72 mph or 63 knots. That is the minimum! The maximum winds a hurricane can generate have never been recorded. The recording instruments and the structure holding them has always blown away first. However, computations based on damage done indicate that winds must have reached velocities between 200 and 250 mph. And to make it worse, these winds are not steady; they are very gusty. A hurricane with winds of 100 knots near the eye may have gusts to 150.

The waves whipped up by these winds will range from 35 to 40 feet in height, and may be higher in particularly severe storms. Long, heavy swells are often the first visible sign well in advance of an oncoming hurricane. The next sign will be thunderstorms that form a band several hundred miles ahead of the main storm.

The normal hurricane cloud sequence starts with high wispy cirrus clouds (Fig. 12-14). These are followed by cirrostratus which gradually lower and change to altostratus. The altostratus become mixed with altocumulus followed by cumulus with very tall vertical development and heavy rain squalls. The barometer is now falling rapidly, and the wind increasing. The heart of the storm is heralded by the approach of a huge, ominous black wall of clouds, a torrential downpour of rain, fierce wind gusts, frequent lightning flashes, and sharply dropping barometer.

When facing the direction toward which the storm is moving, the right side (Fig. 12-15) is called the "dangerous semicircle" (as if the whole thing weren't dangerous!). The left side is optimistically called the "navigable semicircle." The difference between the two sides is that wind speeds on the "dangerous" right side are increased by the speed of the storm's advance, while those on the left are reduced by the same amount.

HURRICANE TRACKS

Most Atlantic / West Indian hurricanes form between about latitude 10 N and 15 N anywhere along the belt from about the Cape Verde Islands to the western Carribean. They start out deeply embedded in the easterly trade winds and move initially toward the west along with those winds. As they move westward, they also work northward finally passing through the horse altitudes to come under the influence of the prevailing westerlies. Embedded now in the westerlies they "recurve" toward the east.

If there is one thing that is consistent about hurricane tracks it is

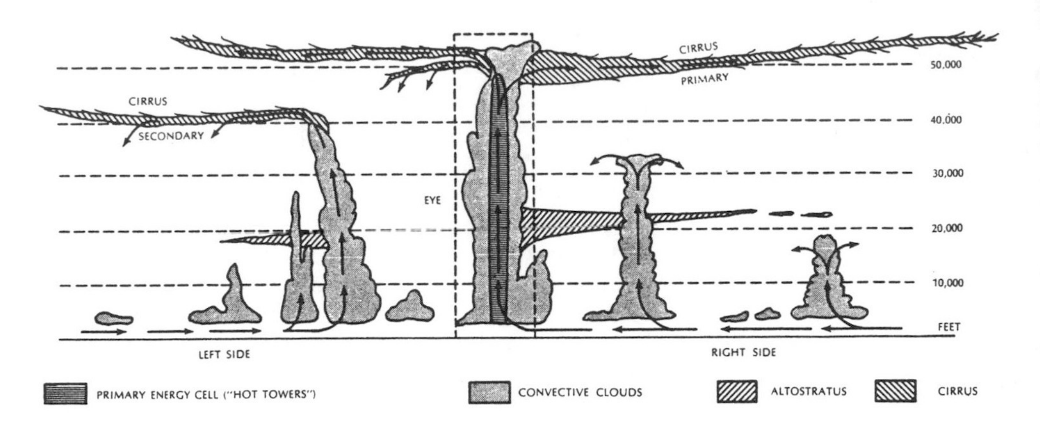

Figure 12-14. Vertical section through a fully developed hurricane. From: Kotsch, *Weather for the Mariner.*

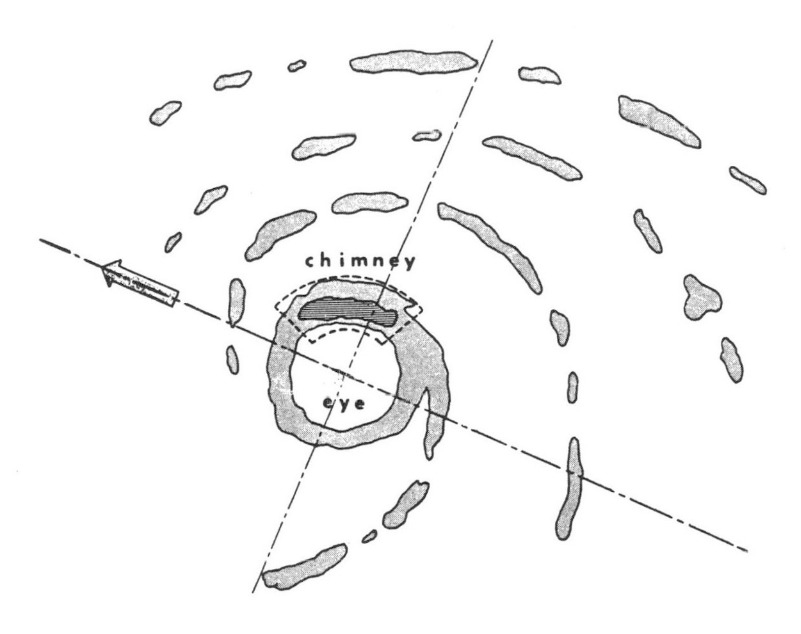

Figure 12-15. Radar view of hurricane showing the eye surrounded by spiraling bands of clouds. From: Kotsch, *Weather for the Mariner.*

that they are erratic. A look at the tracks recorded over a number of years for hurricanes that occurred only during the month of September (Fig. 12-16) in the Atlantic / West Indies area clearly demonstrates this. It also shows that nowhere along either the U.S. East or Gulf coasts is safe from being struck by a hurricane.

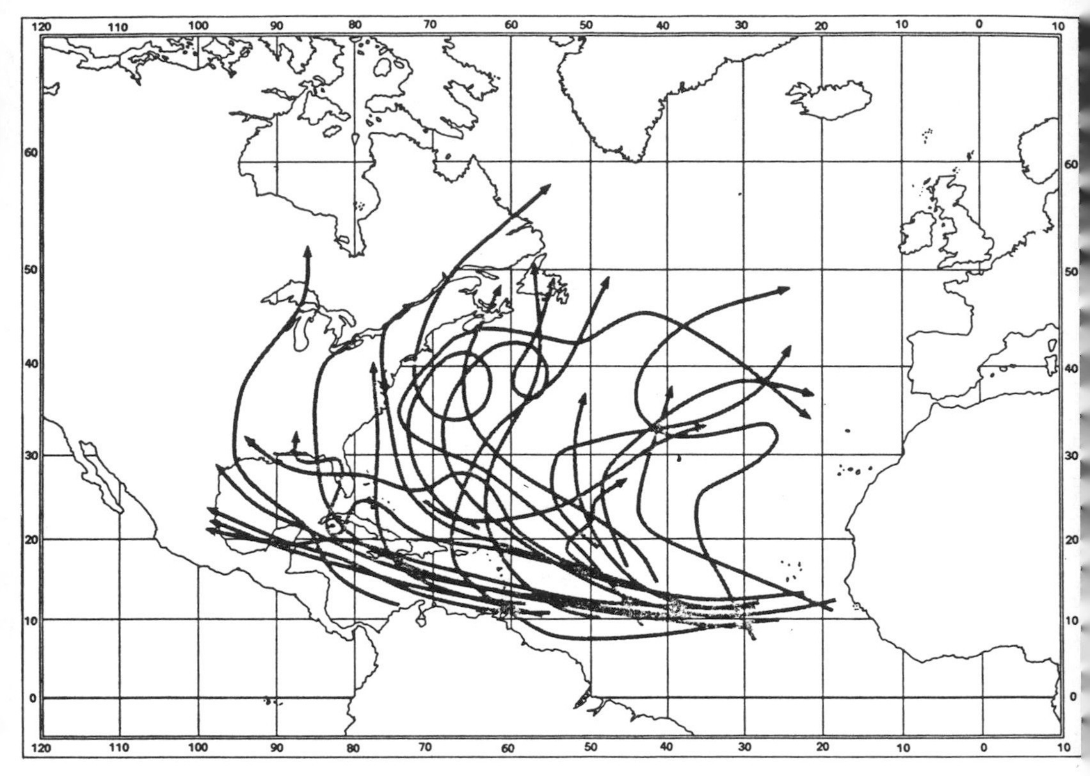

Figure 12-16. Tracks of various hurricanes that have occurred in the month of September showing erratic nature of the tracks. From: *Heavy Weather Guide,* U.S. Naval Institute, 1965

WHAT TO DO

Should you be unfortunate enough to find yourself in the path of a hurricane you can locate yourself relative to the center or eye as follows:

If the wind is gradually veering (changing clockwise) you are in the dangerous semicircle.

If the wind is gradually backing (changing counterclockwise) you are in the navigable semicircle.

If the wind direction is steady with velocity increasing and barometer dropping you are directly in the path of the eye and it is approaching.

If the wind direction is steady with velocity dropping and barometer rising you are safe directly behind the eye.

If unavoidably caught at sea within the area of a hurricane:

IN THE DANGEROUS SEMICIRCLE: take the wind broad on the starboard bow (045 degrees relative) and make the best speed possible.

IN THE NAVIGABLE SEMICIRCLE: take the wind broad on the starboard quarter (135 degrees relative) and make the best speed possible.

IN THE PATH AHEAD OF THE EYE: take the wind aft the starboard quarter (160 degrees relative) and make best speed well into navigable semicircle. Then change to take wind at 135 degrees relative and continue.

IN THE PATH PAST THE EYE: take the most comfortable course that keeps you moving away from the eye.

If you are in port and docked:

Double up all docking lines and put on charfing gear, lay out all the bumpers you can get, take off and stow all sails, canvas covers, dodgers, and anything else the wind can get hold of. If there are any large cabin windows board them up. Secure all hatches and doorways. Help secure any nearby boats that could break loose and hit you. Get well away from the port area until the storm has passed.

If you are in port at anchor:

Set out your heaviest bower anchor with as long a scope as possible. Set a second heavy "underfoot" anchor with short scope so as to drag as the wind veers or backs. This will minimize swinging. Set chafing gear on both anchor lines. As above, remove and stow everything the wind can get at, secure all hatches and doorways, and board up any windows.

SUMMARY

WEATHER SIGNS AND FORECASTING HINTS

We are now familiar with some of the mechanisms that produce weather and weather changes. We are also familiar with a number of the signs to watch for as indicators of what is in store. Keeping a Weather Log (Appendix IV) while aboard, as mentioned in Chapter 3, is probably the best way to practice observing weather signs. In addition, with this written record for reference you can trace changes in clouds, wind, temperature, and atmospheric pressure, so as to interpret the significance of these changes in the light of what you now know about the workings of the atmosphere.

When using your observations to supplement reports from the Weather Service, you will find that in some cases you see exactly what you were told to expect. However, occasionally what you will see seems to be in total disagreement. In such cases maximum safety and minimum discomfort generally lie in accepting the most pessimistic interpretation of the two.

For example, when Weather Service says, "Strong winds and rain." while you are seeing light winds and clear, it is best to assume they know something you don't, and prepare for foul weather anyway. On the other hand, when they are saying, "Fair, winds west to northwest 10 to 12 knots," and you see stratus clouds lowering and thickening with winds freshening out of the southeast, assume that you know something *they* don't, and act accordingly. Remember, there is also a lively possibility that they know about the change all right, but from where they are to where you are there is a lag in communications. The worst that can happen by following this philosophy is that occasionally you will prepare for a spell of foul weather that doesn't occur.

WEATHER SIGNS

Weather signs are of value in forecasting to the extent that you understand the atmospheric conditions they indicate as discussed in previous chapters. For example the old saying *"Halo round the moon, foul weather soon"* was noted early in Chapter 1. We now know that a halo around the moon would be caused by a sheet of altostratus clouds. Either sun or moon appears as a fuzzy ball through such clouds. When a warm front is moving in we know altostratus is followed by continually lowering stratus and finally rain. Thus there is truth in the old saying even though those who originally coined it did not know the mechanisms that caused the foul weather.

Look now at a summary of a few weather signs that have been mentioned earlier in the book, and think of them in terms of some of the atomspheric conditions with which they may be associated.

PRESENT GOOD WEATHER LIKELY TO CONTINUE

SCATTERED SMALL CUMULUS CLOUDS (cool high pressure air mass, or clear warm sector of a cyclone)

WIND LIGHT TO MODERATE FROM WEST TO NORTHWEST (normal undisturbed temperate zone prevailing wind)

BAROMETER STEADY OR RISING SLOWLY (with SW to NW wind a fair weather anticyclone is passing)

NIGHT CLEAR AND COLD (cool high pressure air mass)

CLOUDS BECOMING THINNER AND HIGHER (cool dry air moving under warmer moist air)

TEMPERATURE USUAL FOR THE SEASON (no weather disturbance near)

NIGHT GROUND FOG CLEARS BY MID-MORNING (in clear weather the absence of cloud cover causes the ground to radiate heat quickly at night, and pick it up quickly in morning)

HEAVY NIGHT DEW OR FROST (surface heat radiated through clear sky causes surface condensation of water vapor)

STARS OR MOON BRIGHT AND CLEAR (dry stable air mass)

DAY SEA BREEZE, NIGHT LAND BREEZE (no active weather system in the vicinity)

PRESENT GOOD WEATHER WILL TURN POOR

STRATUS CLOUDS MOVE IN UNDER CIRRUS, LOWERING AND THICKENING (warm front disturbance coming)

CUMULUS CLOUDS GROWING LARGER AND TALLER (cold front or thunderstorm coming)

BAROMETER FALLING STEADILY (low pressure disturbance coming)

SUN BECOMES FUZZY DISC, MOON HALOED OR FUZZY (altostratus clouds preceding warm front)

WIND BACKING (moving counterclockwise SW to S to E—likely cyclonic disturbance)

CLOUDS THICKENING, WIND INCREASING (frontal disturbance coming)

TEMPERATURE UNUSUAL (TOO HIGH OR TOO LOW) FOR SEASON (too high—warm frontal disturbance, too low—cold frontal disturbance)

LINE OF HEAVY, DARK CLOUDS TO THE WEST (cold front moving in)

PRESENT POOR WEATHER WILL IMPROVE

CLOUDS MOVING HIGHER AND BECOMING THINNER (frontal disturbance passing)

WIND VEERING (moving clockwise E to S to SW, warm sector bringing clearing—SW to NW cold front bringing clearing)

BAROMETER RISING (low pressure disturbance has passed)

AFTER HEAVY RAIN WIND MOVES NORTHWEST AND TEMPERATURE DROPS (cold front passing, clearing will follow)

AFTER RAIN WIND MOVES SOUTHWEST AND TEMPERATURE RISES (warm sector of cyclone moving in—clearing will later be followed by cold front disturbance)

STRATUS CLOUDS CHANGING TO CUMULUS TYPE (occluded front passing)

STEADY RAIN CHANGES TO HEAVY INTERMITTENT SHOWERS (occlusion has passed—clearing will follow)

A COLD FRONT HAS PASSED (final portion of an atmospheric disturbance has moved on—clearing will follow)

CROSSED-WIND RULES

Stand with your back to the surface wind; when the upper winds (clouds) come from the left weather will usually worsen.

When the upper winds (clouds) come from the right weather will usually improve.

With upper and lower winds in the same direction present weather will hold

TEMPERATURE WILL MOVE LOWER

WIND SHIFTS FROM SOUTHWEST TO NORTHWEST (cold front passing)

NIGHT SKY IS CLEAR OR SCATTERED CUMULUS ONLY (surface heat radiates rapidly through clear night sky)

COLD FRONT HAS PASSED AND BAROMETER RISING (cold front marks start of cooler air mass)

TEMPERATURE WILL MOVE HIGHER

WIND BACKS FROM NORTHWEST TOWARD SOUTH (front of warm air mass coming)

WARM FRONT HAS PASSED (warm air mass has arrived)

SKY BECOMES OVERCAST AT NIGHT (overcast sky traps heat—"greenhouse effects")

RAIN OR SNOW LIKELY

When a front (warm, cold, or occluded) is approaching.

When the wind is backing.

When the barometer is steadily falling.

Between 1 and 2 days after cirrus clouds in advance of warm front are first seen to lower and thicken.

Between ½ day and 1 day after cirrostratus clouds produce a halo around sun or moon.

After increasing southerly surface winds with clouds above coming from the west.

Within 6 to 8 hours if early morning temperature is high, air is humid, and towering cumulus clouds are forming.

Within the hour on hot summer afternoon, static on the radio, and towering cumulus clouds in sight.

A single one of any of the indicators noted above is never an adequate basis for a weather prediction. You need at least two, or far better three or more indicators all corroborating each other. Your probability of accuracy then becomes quite good.

It is when some of these indicators conflict that weather prediction changes from pure science to an art. Knowledge of the mechanics of the atmosphere has to be coupled with local knowledge, past experience, a hunch, a feeling, guesswork, and a little pure luck.

Sophisticated computer models have vastly improved the accuracy of the Weather Service predictions, both long range and short range. However, in the foreseeable future while advanced computers and instrumentation will greatly expand our information gathering capability, they will in no way replace the experienced human eye and mind when it comes to interpreting that information.

Motorboatman, sailor, canoeist, or fisherman, all us who work or play on the water need all the weather information we can get in order to maintain both our safety and our comfort. Let us close then with the optimistic thought that for the sailor maximum safely lies in collecting all the information possible and then taking the most pessimistic interpretation of all the available signs.

BIBLIOGRAPHY

Beyers, Horace R. *General Meteorology*. New York: McGraw-Hill Book Co., 1959.

Blair and Fite. *Weather Elements*. Englewood Cliffs, N.J.: Prentice-Hall, Inc., 1965.

Bowditch, N. *American Practical Navigator* Washington, D.C.: U.S. Government Printing Office, 1977.

Colon, Jose A. *Climatologica de Puerto Rico*. San Juan, P.R.: National Weather Service.

Donn, William L. *Meteorology* New York: McGraw-Hill Book Co., 1965.

Figgins, W. E. *Climate of Salt Lake City Utah*. U.S. Weather Service, 1980.

Gross, M. Grant. *Oceanography*. Englewook Cliffs, N.J.: Prentice-Hall, Inc., 1977.

Harding and Kotsch. *Heavy Weather Guide*. Annapolis, M.D.: Naval Institute Press, 1965.

Kotsch, William J. *Weather for the Mariner*. Annapolis, M.D.: Naval Institute Press, 1983.

Kurtz, Emil S. *Southern California Weather for Small Boaters,* Salt Lake City, U.T.: National Weather Service, 1971.

Maloney, Elbert S. *Dutton's Navigation and Piloting*. Annapolis, M.D.: Naval Institute Press, 1978.

Petterssen, Sverre. *Introduction to Meteorology* New York: McGraw-Hill Book Co., 1969.

Watts, Allen. *Instant Weather Forecasting*. New York: Dodd, Mead & Co., 1968.

FAA and NOAA. *Aviation Weather*. Washington, D.C.: National Weather Service, 1975.

Hardy, Wright, Kington, and Gribbin. *The Weather Book*. Boston: Little Brown & Co., 1982.

U.S. Coast Pilot 1. National Oceanographic and Atmospheric Administration Washington, D.C., 1985.

U.S. Coast Pilot 2. NOAA. Washington, D.C., 1985.

U.S. Coast Pilot 3. NOAA. Washington, D.C., 1984.

U.S. Coast Pilot 4. NOAA. Washington, D.C., 1985.

U.S. Coast Pilot 5. NOAA. Washington, D.C., 1984.

U.S. Coast Pilot 6. NOAA. Washington, D.C., 1985.

U.S. Coast Pilot 7. NOAA. Washington, D.C., 1971.

Hurricane. NOAA. Washington, D.C., 1971.

APPENDICES

TEMPERATURE

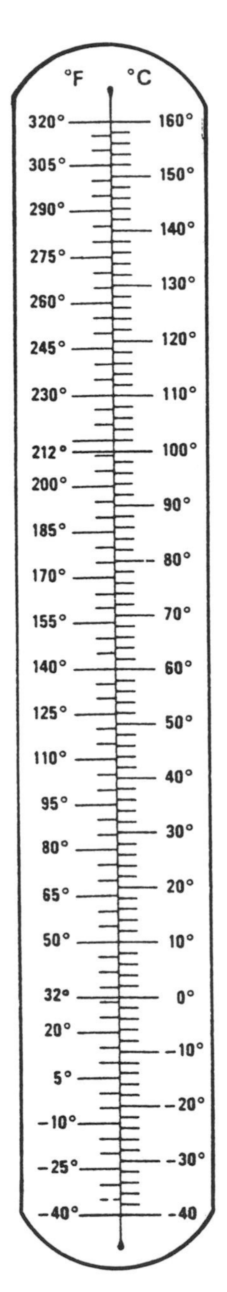

32° Fahrenheit = 0° Celsius
0° Fahrenheit = − 17.8° Celsius

Appendix I. Fahrenheit to centigrade conversion scale.

TABLE A

Inches of Mercury	*Millibars*	*Inches of Mercury*	*Millibars*	*Inches of Mercury*	*Millibars*
1	33.86	11	372.50	21	711.14
2	67.73	12	406.37	22	745.01
3	101.59	13	440.23	23	778.87
4	135.46	14	474.09	24	812.73
5	169.32	15	507.96	25	846.60
6	203.18	16	541.82	26	880.46
7	237.05	17	575.69	27	914.33
8	270.91	18	609.55	28	948.19
9	304.78	19	643.41	29	982.05
10	338.64	20	677.28	30	1015.92
				31	1049.78
				32	1083.65

TABLE B

Inches of Mercury	**0.00**	**.01**	**.02**	**.03**	**.04**	**.05**	**.06**	**.07**	**.08**	**.09**
0.0	0.00	0.34	0.68	1.02	1.35	1.69	2.03	2.37	2.71	3.05
0.1	3.39	3.73	4.06	4.40	4.74	5.08	5.42	5.76	6.10	6.43
0.2	6.77	7.11	7.45	7.79	8.13	8.47	8.80	9.14	9.48	9.82
0.3	10.16	10.50	10.84	11.18	11.51	11.85	12.19	12.53	12.87	13.21
0.4	13.55	13.88	14.22	14.56	14.90	15.24	15.58	15.92	16.25	16.59
0.5	16.93	17.27	17.61	17.95	18.29	18.63	18.96	19.30	19.64	19.98
0.6	20.32	20.66	21.00	21.33	21.67	22.01	22.35	22.69	23.03	23.37
0.7	23.70	24.04	24.38	24.72	25.06	25.40	25.74	26.08	26.41	26.75
0.8	27.09	27.43	27.77	28.11	28.45	28.78	29.12	29.46	29.80	30.14
0.9	30.48	30.82	31.15	31.49	31.83	32.17	32.51	32.85	33.19	33.53

Table A converts inches of mercury in round numbers to millibars. Table B converts inches of mercury in tenths and hundredths to millibars. Both tables can be used together to solve a conversion problem.

Example: Convert 29.73 inches of mercury to millibars.

Solution: First determine from table A the number of millibars in 29 inches of mercury.

Then determine from table B the number of millibars in .73 inches of mercury. In table B the vertical axis lists inches of mercury in tenths of an inch. The horizontal axis lists inches of mercury in hundredths of an inch. To convert .73 inches, read down the vertical axis to .7 and across the horizontal axis to .03.

Now add the data from both tables for the answer.

From table A	982.05 mbs
From table B	24.72 mbs
Answer:	1006.77 mbs

Appendix II. Conversion tables—inches of mercury to millibars. From: Kotsch, *Weather for the Mariner.*

RELATIVE HUMIDITY TABLE

AIR TEMP., °F.	DEPRESSION OF WET-BULB THERMOMETER, °F.															
	1	2	3	4	5	6	7	8	9	10	11	12	13	14	15	16
0	67	33	1													
5	73	46	20													
10	78	56	34	13												
15	82	64	46	29	11											
20	85	70	55	40	26	12										
25	87	74	62	49	37	25	13	1								
30	89	78	67	56	46	36	26	16	6							
35	91	81	72	63	54	45	36	27	19	10	2					
40	92	83	75	68	60	52	45	37	29	22	15	7				
45	93	86	78	71	64	57	51	44	38	31	25	18	12	6		
50	93	87	80	74	67	61	55	49	43	38	32	27	21	16	10	5
55	94	88	82	76	70	65	59	54	49	43	38	33	28	23	19	11
60	94	89	83	78	73	68	63	58	53	48	43	39	34	30	26	21
65	95	90	85	80	75	70	66	61	56	52	48	44	39	35	31	27
70	95	90	86	81	77	72	68	64	59	55	51	48	44	40	36	33
75	96	91	86	82	78	74	70	66	62	58	54	51	47	44	40	37
80	96	91	87	83	79	75	72	68	64	61	57	54	50	47	44	41
85	96	92	88	84	81	77	73	70	66	63	59	57	53	50	47	44
90	96	92	89	85	81	78	74	71	68	65	61	58	55	52	49	47
95	96	93	89	86	82	79	76	73	69	66	63	61	58	55	52	50
100	96	93	89	86	83	80	77	73	70	68	65	62	59	56	54	51
105	97	93	90	87	84	81	78	75	72	69	66	64	61	58	56	53
110	97	93	90	87	84	81	78	75	73	70	67	65	62	60	57	55
115	97	94	91	88	85	82	79	76	74	71	69	66	64	61	59	57
120	97	94	91	88	85	82	80	77	74	72	69	67	65	62	60	58
125	97	94	91	88	86	83	80	78	75	73	70	68	66	64	61	59
130	97	94	91	89	86	83	81	78	76	73	71	69	67	64	62	60

Appendix III. Relative humidity table for use with sling psychrometer. Relative humidity is given in percentage.

17	18	19	20	21	22	23	24	25	26	27	28	29	30	31	32	33	34	35
9	5																	
17	13	9	5	1														
24	20	16	12	9	5	2												
29	25	22	19	15	12	9	6	3										
34	30	27	24	21	18	15	12	9	7	4	1							
38	35	32	29	26	23	20	18	15	12	10	7	5	3					
41	38	36	33	30	27	25	22	20	17	15	13	10	8	6	4	2		
44	41	39	36	34	31	29	26	24	22	19	17	15	13	11	9	7	5	3
47	44	42	39	37	34	32	30	28	25	23	21	19	17	15	13	11	10	8
49	46	44	41	39	37	35	33	30	28	26	24	22	21	19	17	15	13	12
51	49	46	44	42	40	38	36	34	32	30	28	26	24	22	21	19	17	15
52	50	48	46	44	42	40	38	36	34	32	30	28	26	25	23	21	20	18
54	52	50	48	46	44	42	40	38	36	34	33	31	29	28	26	25	23	21
55	53	51	49	47	45	43	41	40	38	36	34	33	31	29	28	26	25	23
57	55	53	51	49	47	45	44	42	40	38	37	35	33	32	30	29	27	26
58	56	54	52	50	48	47	45	43	41	40	38	37	35	33	32	30	29	28

WEATHER LOG

DATE AND TIME	TEMP.		REL. HUM.	DEW POINT	PRESSURE	
	Dry Bulb	*Wet Bulb*			*Bar.*	*Tend.*

Appendix IV. Sample blank page for weather log.

WIND		VIS.	CLOUDS		WEATHER AND COMMENTS
Dir.	*Vel.*		*Type*	%	

TABLE OF DEW POINT TEMPERATURES—*for use with sling psychrometer*

AIR TEMP., °F.	VAPOR PRESSURE	DEPRESSION OF WET-BULB THERMOMETER, °F.														
		1	2	3	4	5	6	7	8	9	10	11	12	13	14	15
0	0.0383	−7	−20													
5	0.0491	−1	−9	−24												
10	0.0631	5	−2	−10	−27											
15	0.0810	11	6	0	−9	−26										
20	0.103	16	12	8	2	−7	−21									
25	0.130	22	19	15	10	5	−3	−15	−51							
30	0.164	27	25	21	18	14	8	2	−7	−25						
35	0.203	33	30	28	25	21	17	13	7	0	−11	−41				
40	0.247	38	35	33	30	28	25	21	18	13	7	−1	−14			
45	0.298	43	41	38	36	34	31	28	25	22	18	13	7	−1	14	
50	0.360	48	46	44	42	40	37	34	32	29	26	22	18	13	8	[illegible]
55	0.432	53	51	50	48	45	43	41	38	36	33	30	27	24	20	15
60	0.517	58	57	55	53	51	49	47	45	43	40	38	35	32	29	25
65	0.616	63	62	60	59	57	55	53	51	49	47	45	42	40	37	34
70	0.732	69	67	65	64	62	61	59	57	55	53	51	49	47	44	42
75	0.866	74	72	71	69	68	66	64	63	61	59	57	55	54	51	49
80	1.022	79	77	76	74	73	72	70	68	67	65	63	62	60	58	56
85	1.201	84	82	81	80	78	77	75	74	72	71	69	68	66	64	62
90	1.408	89	87	86	85	83	82	81	79	78	76	75	73	72	70	69
95	1.645	94	93	91	90	89	87	86	85	83	82	80	79	78	76	74
100	1.916	99	98	96	95	94	93	91	90	89	87	86	85	83	82	80
105	2.225	104	103	101	100	99	98	96	95	94	93	91	90	89	87	86
110	2.576	109	108	106	105	104	103	102	100	99	98	97	95	94	93	91
115	2.975	114	113	112	110	109	108	107	106	104	103	102	101	99	98	97
120	3.425	119	118	117	115	114	113	112	111	110	108	107	106	105	104	102
125	3.933	124	123	122	121	119	118	117	116	115	114	112	111	110	109	108
130	4.504	129	128	127	126	124	123	122	121	120	119	118	116	115	114	113

Appendix VI. Table of dew point temperatures, for use with sling psychrometer.

	17	18	19	20	21	22	23	24	25	26	27	28	29	30	31	32	33	34	35
3																			
9	1	−12	−59																
1	17	11	4	−8	−36														
1	27	24	19	14	7	−3	−22												
9	36	33	30	26	22	17	11	2	−11										
7	44	42	39	36	32	29	25	21	15	8	−2	−23							
4	52	50	47	44	42	39	36	32	28	24	20	13	6	−7	−53				
1	59	57	54	52	50	48	45	42	39	36	32	28	24	19	12	3	−12		
7	65	63	61	59	57	55	53	51	48	45	43	39	36	32	28	24	19	11	1
3	71	70	68	66	64	62	60	58	56	54	52	49	46	43	40	37	33	29	24
9	77	76	74	72	71	69	67	65	63	61	59	57	55	52	50	47	44	41	37
4	83	82	80	78	77	75	74	72	70	68	67	65	63	61	58	56	54	51	48
0	89	87	86	84	83	81	80	78	77	75	73	72	70	68	66	64	62	60	57
6	94	93	92	90	89	87	86	84	83	81	80	78	76	75	73	71	69	67	65
1	100	98	97	96	94	93	92	90	89	87	86	84	83	81	80	78	76	75	73
6	105	104	103	101	100	99	97	96	95	93	92	90	89	88	86	84	83	81	80
2	110	109	108	107	106	104	103	102	100	99	98	96	95	94	92	91	89	88	86

WIND BAROMETER TABLE

WIND DIRECTION	BAROMETER REDUCED TO SEA LEVEL	CHARACTER OF WEATHER
SW to NW	30.10 to 30.20 and steady	Fair, with slight temperature changes for one or two days.
SW to NW	30.10 to 30.20 and rising rapidly	Fair followed within two days by rain.
SW to NW	30.20 and above and stationary	Continued fair with no decided temperature change.
SW to NW	30.20 and above and falling slowly	Slowly rising temperature and fair for two days.
S to SE	30.10 to 30.20 and falling slowly	Rain within twenty-four hours.
S to SE	30.10 to 30.20 and falling rapidly	Wind increasing in force, with rain within 12 to 24 hours.
SE to NE	30.10 to 30.20 and falling slowly	Rain in 12 to 18 hours.
SE to NE	30.10 to 30.20 and falling rapidly	Increasing wind and rain within 12 hours.
E to NE	30.10 and above and falling slowly	In summer with light winds rain may not fall for several days. In winter rain in 24 hours.
E to NE	30.10 and above and falling fast	In summer, rain probably in 12 hours. In winter, rain or snow, with increasing winds will often set in when the barometer begins to fall and the wind set in NE.
SE to NE	30.00 or below and falling slowly	Rain will continue 1 or 2 days.
SE to NE	30.00 or below and falling rapidly	Rain with high wind, followed within 36 hours by clearing and in winter colder.
S to SW	30.00 or below and rising slowly	Clearing in a few hours and fair for several days.
S to E	29.80 or below and falling rapidly	Severe storm imminent, followed in 24 hours by clearing and in winter colder.
E to N	29.80 or below and falling rapidly	Severe NE gale and heavy rain, winter heavy snow and cold wave.
Going to W	29.80 or below and rising rapidly	Clearing and colder.

Reproduced by Courtesy of National Weather Service

Appendix VI. Wind barometer table. Courtesy: National Weather Service.

INDEX